TIME WARPED

Also by Claudia Hammond

Emotional Rollercoaster:
A Journey Through the Science of Feelings

TIME WARPED

Unlocking the Mysteries
of Time Perception

CLAUDIA HAMMOND

HARPER ● PERENNIAL

NEW YORK ● LONDON ● TORONTO ● SYDNEY ● NEW DELHI ● AUCKLAND

HARPER  PERENNIAL

Originally published in Great Britain in 2012 by Canongate Books Ltd.

FIRST U.S. EDITION

Library of Congress Cataloging-in-Publication Data is available upon request.

ISBN 978-0-06-222520-7

13 14 15 16 17   /RRD   10 9 8 7 6 5 4 3 2 1

For Tim

The only reason for time is so that everything doesn't happen at once.

Albert Einstein

CONTENTS

The River of Time – Making Time Go Backwards – Mellow Monday and Furious Friday

Autobiographical Memory – Total Recall – When Time Speeds Up – Life Through a Telescope – Take Two Items a Day For Five Years – Time-Stamping the Past – Everything Shook – A Thousand Days – The Reminiscence Bump – Remembering Moments, Not Days – The Holiday Paradox

Time-Travelling Into the Future – Can Your Dog Picture Next Week? – What Are You Doing Tomorrow? – Memories For Events That Never Happened – Suicide Island – Thinking About Nothing – An Erroneous Future – Bad Choices – Five Years to Reach the Word 'Ant' – One Marshmallow Or Two? – Future-Orientated Thinking – Looking Back, Looking Forward

Time Is Speeding Up – Making Time Go Faster – Too Much To Do, Too Little Time – Failing to Plan Ahead – A Poor Memory For the Past – Worrying Too Much About the Future – Trying to Live in the Present – Predicting How You'll Feel in the Future – In Conclusion

INTRODUCTION

WHEN CHUCK BERRY finds himself at the edge of a cliff or at the top of a mountain, he likes to jump off. When he's in a plane, he likes to jump out. This is not Chuck Berry the famous rock-and-roll singer, I hasten to add, but Chuck Berry 'the Kiwi king of skydiving and base-jumping'. You may well have seen him in adverts for fizzy drinks. For Lilt, he jumped out of a helicopter while riding a bicycle – twice. Now he's sponsored by Red Bull, but you can be sure that he experiences more than a caffeine rush as he falls through the air with a parachute, choosing not to open it until the last possible moment.

For 25 years Chuck Berry has practised plummeting through the sky, whether skydiving, hang-gliding, micro-light-flying or parachuting (once he even used a customised tent as a canopy); but his speciality is base-jumping. One of the more extreme 'extreme sports', it takes its name from the four categories of fixed objects from which you can jump – buildings, antennae, spans (bridges) and the earth (in practice, a cliff). Since 1981 there have been at least

136 fatalities; it is a sport so dangerous that one in 60 participants is expected to die doing it.

For Chuck, the key to survival resides in his ability to control his mind. Before he leaps, he visualises the exact steps he will take to achieve a successful outcome. So while any of the rest of us who found ourselves teetering on the top of the world's tallest building (the K.L. Tower in Kuala Lumpur) would be likely to picture all the things that could go wrong – getting blown into another building, opening the parachute too late, ending up a bloody mess on the street 1,381 feet below – Chuck carefully calculates the wind direction, decides on the optimal point to open his chute and pictures himself floating down to make the perfect landing on the selected spot. Of course it helps that he also does months of planning.

With so many years' experience, a Swift flight that Chuck took one New Year's Day should have been easy. A Swift is a cross between a plane and a hang-glider and is said to combine the glorious soaring abilities of a glider with the convenience of being able to ascend into the air by simply running off the side of a mountain – no need for a plane to tow you up into the sky. What's more it folds up small enough to fit on top of a roof rack. The front half looks like an elegant paper plane with extra-long, aerodynamic wings, while the body of plane is very short and the tail is missing altogether. There's a little cockpit for the pilot, which covers only your head, shoulders and arms, and your legs hang out of the bottom to allow you to run down the hill. Picture Fred Flintstone running along the ground to start his Stone Age car – and then disappearing over the edge of the cliff and taking flight.

For his flight in the Swift, Chuck chose Coronet Peak just outside New Zealand's bungee-jumping capital, Queenstown. It was a beautiful summer's day and the mountain was outlined against the deep blue sky like a theatre scenery flat. It should have been the perfect location, but for Chuck the idea of some gentle soaring in this awesome immensity was a little tame. Some aerial acrobatics would sharpen the thrill. So, riding a thermal, he took the hybrid glider up to a height of 5,500 feet and then plunged her into a steep nosedive. The plan was to cut the dive at the last possible moment and then bank up towards the heavens again. No problem, right?

Wrong. The whole glider started shaking and bucking violently, and, as a former aircraft engineer, Chuck knew exactly what was happening. This was what was known in the trade as 'flutter' – a term devised by someone big on understatement – when the wings of an aircraft twist up and down repeatedly until eventually they flap themselves to death.

Within moments, both wings had sheared off completely and Chuck found himself in freefall. Speeding towards the ground was usually his idea of fun. But this time there was nothing to slow his descent, nothing to break his fall, nothing to save him hitting the ground at breakneck speed. Even so, as he hurtled to earth – his GPS tracker would later show rescuers that he had fallen at a speed of 200km an hour – Chuck's mind was capable of detailed, rational thought.

Although he was now hanging outside the cockpit of a glider without wings, looking up, he saw that he was still attached to much of the wreckage. His mind went into overdrive. He remembers exactly what he was thinking:

There had to be a way to get back into the remains of the glider. Why couldn't he climb up into the cockpit? There had to be a way. Can't I pull myself up? Surely. What would James Bond do? Come on, dude, do something! I have to do something. Don't look at the ground. It's too close. There's no time. But there has to be a way. It must have been flutter. The lever! The lever for the emergency chute. If I can just get to the lever. It should be there! Surely it has to be there. How long have I been falling for? This is taking ages. There are the hills. Not much time left. Too windy to think. This is the most important deci- sion I'll ever make. Do something! Save yourself! Get to that lever and pull!

Now, bear in mind that this interior dialogue, all of these thought processes, all of this precise mental calculation, would later be revealed by Chuck's GPS system to have taken place within a matter of seconds. But for Chuck it felt like a lot longer. He knew that he had to act fast, but he had enough time – plenty it seemed – to think and to take action. For the observer, the seconds flashed by. For Chuck, it seemed to extend almost endlessly. The same time-frame with two very different perceptions of time passing. His New Year's Day glimpse of eternity is a perfect, if extreme, example of this book's central theme: the subjec- tivity of the experience of time. In situations like the one he faced, time is weirdly elastic.

We have all experienced moments in life where time becomes warped. When we are in fear of our lives, like Chuck, it seems to slow down. When we are enjoying ourselves, time 'flies'. As the years go by, life feels as though it's speeding up. Christmas comes round that bit faster every year. Yet when we were children, the school holidays seemed to stretch on for months.

In this book I'll be asking whether this stretching and shrinking of time is simply an illusion or whether the mind processes time differently at different moments of our lives. Time perception – the way we subjectively experience time, what time feels like to us as individuals – is an endlessly fascinating topic because time constantly surprises us; we never quite get used to the way it plays tricks on us. A good holiday races by; no sooner have you settled in than it's time to think about packing to leave. Yet the moment you arrive home, it feels as though you have been away ages. How is it possible to have such contradictory experiences of the same holiday?

At the core of this book is the idea that the experience of time is actively *created* by our minds. Various factors are crucial to this construction of the perception of time – memory, concentration, emotion and the sense we have that time is somehow rooted in space. It's this last factor that allows us to do something extraordinary – to time-travel at will in our minds, moving backwards and forwards through time. I will focus on psychology and brain science, rather than the metaphysics and poetry of time, or the physics and philosophy of time – although sometimes it is hard to know where one field ends and another begins.

Physicists tell us that the popular notion that time is segmented into the past, the present and the future is inaccurate. Time does not pass; time simply *is*. John Ellis McTaggart, a well-known philosopher of time, believed much the same thing,[1] and versions of this idea underpin religions such as Buddhism and Hinduism. But this book isn't about the objective reality of time, but rather the *experience* of it, and I'm confident that you, like me, experience time as flow, rather than stasis. I'll be concentrating on how the mind creates sensations of time; the time that neuroscientists and psychologists call 'mind time'. This is a time that can't be measured by an external clock, but is central to our experience of reality.

I shall be revealing some of the imaginative methods that researchers in the emerging field of the psychology of time have used to study mind time. They've quizzed people on the dates of famous events, had them steer themselves towards precipices and even thrown them backwards off buildings. They haven't been afraid to experiment on themselves either – spending months living alone in an ice cave without daylight, or measuring their own time estimation skills every single day for 45 years. Then there are those who have experienced events that have unintentionally revealed a great deal about time perception, like the man who lost the ability to imagine the future after a motorbike accident, and the BBC journalist who spent more than three months as a hostage without knowing whether he would ever be released.

Combining these experiences with cutting edge research in psychology and neuroscience from around the world gives us an invaluable insight into the curious nature of time perception. We all know something of the malleability of time, and we don't have to go to Chuck's extremes to experience it. Psychologists have discovered some extraordinary things: among them the fact that eating fast food makes us feel impatient,[2] the fact that people at the back of a queue are more likely to see time as moving towards them, while people at the front see themselves as moving through time; the fact that if someone has a raging temperature time goes more slowly.

There's also my own theory of the 'Holiday Paradox', which explains the phenomenon I referred to where holidays pass quickly, yet feel as though they lasted a long time afterwards. We constantly observe time in two ways – while it's happening and then in retrospect. Most of the time this dual viewpoint serves us well, but it provides the key to many of the mysteries of time. When the two perceptions – prospective and retrospective – fail to tally, time feels confusing.

I will be revealing the results of my study into the way people visualise time in their minds. You may be surprised to learn that one in five of us imagine the days, the months, the years and even the centuries laid out in precise patterns in the mind's eye. The variety in the way people visualise time is intriguing too – with centuries standing like dominoes or decades shaped like a slinky. Why do some people see time like this and what effect does it have on their experience of time? And I'll be addressing a question that has no right or wrong answer but still divides us – is the future coming towards us or are we endlessly moving along a timeline towards the future?

Today we can calculate the time more precisely, more
minutely, than ever before. The caesium clock at the
National Institute of Standards and Technology in the
United States is so accurate that in the next 60 million years
it will neither gain, nor lose, a second. A few years ago it
could only do that for a mere 20 million years. The clock
of the mind is more elusive. It seems to govern our experience of time, yet appears not to exist. For decades scientists
have searched for evidence of an internal clock. Over a
24-hour period, circadian rhythms regulate our body clock,
keeping us in synch with day and night through exposure
to daylight, but there is no single organ dedicated to sensing
the seconds, minutes or hours passing. Nevertheless our
minds can measure time. We can estimate a minute fairly
accurately. We constantly deal with different time-frames
– a moment ago, middle age, the past decade, the first week
of term, every Christmas, two hours' time – which we juggle
effortlessly in our minds. Meanwhile we are building up a
long-term sense of the decades passing, and of our own
life history and where we fit into the earth's history.

The latest findings from neuroscience are beginning to
give us clues as to how our brains can sense time without
any single organ devoted to the purpose and in Chapter
Two I will examine these competing neuroscientific theories. But it may be that what fascinates you more is how
your conception of time affects the way you think and the
way you behave. Although according to the calendar time
only goes in one direction, in our minds we constantly leap
about from the past to the future and back again. If you
wish, you can read this book in the same way. While I

think I've set it out in the right order, you don't have to follow me. If you have ever wondered how good you are at making decisions based on how you might feel in the future, Chapter Five beckons. If you've ever been in an accident and experienced time standing still, in Chapter One you can find out why. If you want to know why time feels as though it's speeding up or why you world news events always feel as if they happened a year or two longer ago than you thought, then Chapter Three may be for you.

To conclude, I will explore how all this research might be useful in our everyday lives. We construct the experience of time in our minds, so it follows that we are able to change the elements we find troubling – whether it's trying to stop the years racing past, or speeding up time when we're stuck in a queue, trying to live more in the present, or working out how long ago we last saw our old friends. Time can be a friend, but it can also be an enemy. The trick is to harness it, whether at home, at work, or even in social policy, and to work in line with our conception of time. Time perception matters because it is the experience of time that roots us in our mental reality. Time is not only at the heart of the way we organise life, but the way we experience it.

Finally, a word about the word 'time'. Naturally a book about time will use the word a lot. Were I from the Amondawa tribe in the Amazon that would be a problem. They have no word for time, no word for month and no word for year. There is no agreed calendar and there are no clocks. They do refer to sequences of events, but time does not exist as a separate concept. By contrast, the word 'time' is used more often than any other noun in the English language.[3] This reflects our fascination with time – and is one reason why I've written this book. But the ubiquity of the word presents some difficulties – as it is all too easy to use time all the time. You see the problem? To avoid confusion, I may sometimes seem pedantic about terms or utilise jargon from psychology. There are also phrases, such as future thinking, which, for the sake of precision, at times, I will use repeatedly. I hope you will bear with me.

Now, I'm sure you're wondering what happened to Chuck Berry, our base-jumping glider pilot who was left suspended in the air, his body falling and time dilating. I'm afraid you won't find out right away, as there are many other issues to explore. But at the end of the next chapter, we will use our ability to travel back in time in our minds and we'll learn how it all worked out for Chuck.

THE TIME ILLUSION

WHEN THE BBC reporter Alan Johnston was held captive in Palestinian-controlled Gaza, he had plenty of time to fill but no accurate method of measuring it. With no wristwatch, no books, or pen and paper, his only means of guessing how much time had passed was by studying the lines of light visible through the shutters and the shadow that moved slowly across the walls as he willed each day away. The five Islamic calls to prayer also allowed him to work out the rough time of day, but he soon lost track of the date. 'I made a mark on the door in the traditional clichéd prisoner way, but for a while I was worried about what the guard might do if he saw these marks on the door of his flat. He was going through quite bad moods at the time, so I started making etching lines on the edge of my toothbrush instead, but it was still quite easy to become uncertain about the date and soon I was adrift from time.'

In fact Alan Johnston spent almost four months in that flat, but at the time he had no idea how long he would be detained, or whether he would live or die. 'Suddenly time

becomes like a living thing, a crushing weight that you have
to endure. It's endless, since you don't know when you're
going to be freed, if ever. There's this great sea of time
ahead of you that you have to keep ploughing through.' To
pass the hours, Alan invented mind games, setting himself
tasks such as developing the best possible intellectual attack
on the idea of apartheid, or trying to write poetry and
stories in his head. But with no pen and paper to record
his thoughts, it became an exercise in memory, 'If you write
seven crap lines of poetry, you've got to remember them
before you can move on to the eighth and then when you've
written the ninth line you've got to ask yourself whether
you still remember line five.' Eventually Alan developed his
own mental strategy for coping with the hours, a strategy
that used the concept of time itself – one I'll return to later
in the book.

There were two elements holding sway over Alan's life
as a hostage: his captors and time. In this chapter I'll
examine the conditions under which time can become so
warped that it slows down to the unbearably protracted
pace experienced by Alan Johnston. It is not surprising that
time dragged for him, locked in one room and deprived of
all stimulation, but I'll also be covering other, more peculiar,
circumstances where time expands. It is the mysterious
flexibility of time that makes it so fascinating, but before
we get to that, let's consider why our ability to sense the
passage of time is so important, both to us as individuals
and as part of society.

Accurate timing is essential for communication, co-
operation and human relations in more ways than you might

expect. It's obvious that any activity involving two or more people requires the co-ordination of timetables, but even something as apparently simple as a conversation demands split-second timing. To produce and understand speech, we rely on critical timings of less than a tenth of a second. The difference between the sound of a 'pa' and a 'ba' is all in the timing of the delay before the subsequent vowel, so if the delay is longer you hear a 'p', if it's short you hear a 'b'. If you put your hand on your vocal cords you can even feel that with the 'ba' your lips open at the same time as you feel your cords start to vibrate. With the 'pa' the vibration starts a moment later. This relies on timing accurate to the millisecond. Even the timing between syllables can be crucial to a phrase's meaning. With Jimi Hendrix's lyric, 'Excuse me while I kiss the sky,' just a fraction of a second difference in timing is what gives you the famous mondegreen, 'Excuse me while I kiss this guy.' In order to co-ordinate limb and muscle movements we need to estimate milliseconds, while the appraisal of seconds allows us to do everything from detecting rhythm in music or kicking a ball to deciding whether it's faster to walk along the travelator at the airport or on the floor around the side. (Answer: it depends. Researchers at Princeton University found that taking the travelator usually slows you down because you tend to reduce your pace, or – more irritatingly – get stuck behind people who stop walking as soon as they get on. An empty travelator will get you across the airport faster than walking on the floor alongside it, but only if you don't decide to stand still yourself.)

Our sense of timing isn't perfect, yet on the whole our

brains are able to conceal this, presenting us with a world where time usually feels smooth and consistent. A badly-dubbed film has to be quite bad for us to notice the discrepancy; studies have shown that if the mismatch is anything below 70 milliseconds our brain goes along with our expectation that if we can see a person's mouth moving and we can hear a sound that matches it, then they must be occurring simultaneously. Yet once people are *told* that they don't match, they can then work out whether the pictures are ahead of or behind the sound. So it's not that we *can't* detect these discrepancies, it's that unless we are alerted to a problem the brain assumes that sound and sight fit together because that's what we're used to. Some of our senses are better at timing than others; it's much easier to remember an auditory rhythm tapped out in Morse code than the same series of dots and dashes written down.

The rabbit illusion in the box below is one you can try on someone else.

Find a volunteer, take their forearm and get them to look away. Using the end of a pen, tap very fast several times in the same spot near their wrist and – without breaking the rhythm – tap several times nearer the inside of their elbow. Then ask them what you did.

The chances are they will say that you made a series of taps at regular intervals along their arm from wrist to elbow. Even though you didn't touch the middle of their forearm the brain makes certain assumptions about the distance and the timing of the taps. Likewise if you turn a light on and off very fast it appears to flicker, but if you do it even faster there's a point at which it appears to be perpetually switched on; our brains try to make sense of the flicker by perceiving it as a constant light. We are tagging events in time in order to make sense of them.

Computers with accurate timing in the millisecond range have made it a great deal easier for scientists to investigate the time intervals that the brain can and cannot detect. In the 1880s the Austrian physiologist Sigmund Exner was determined to calculate the shortest time a human could differentiate between two sounds. To do this he used a Savart wheel, a metal disc with teeth all the way round that produce a loud click as the wheel turns. If the wheel goes fast enough then, like the flickering light bulb, the sound appears continuous. Exner wanted to establish the minimum time interval at which humans could still hear separate clicks. He tried the same with electrical sparks and found our senses varied dramatically – when watching sparks, people found it hard to differentiate, but when it came to listening to clicks people could differentiate between two clicks with only a five-hundredth of a second between them.[4]

These are impressive millisecond judgements, but our abilities at time perception go way beyond this. The subjective experience of time relies on the ability to put that millisecond moment into context. As the philosopher Edmund Husserl said in his study of the phenomenology of time, we hear a song one note at a time, but it is our sense of the future and the past – our memory and our anticipation – that makes it a song.[5] The experience of time feels personal, a part of our consciousness that we find hard to put into words. St Augustine wrote 'What then is time? If no one asks me, then I know. If I wish to explain it to someone who asks, I know it not.' Yet we constantly refer to abstract ideas involving time – six months, last week or next year – and everyone knows what we mean. The notion of time is both personal and shared.

YOUR TIME IS MY TIME

Each society forms rules about time that its people share and understand. In many parts of the world, including Europe and the US, if a ticket for a play says 7.30 p.m. it is customary to arrive earlier than the specified time, but if a party invitation says 7.30 p.m. you are expected to arrive later. The sociologist Eviatar Zerubavel believes that these social rules provide us with a means of judging time.[6] We learn to expect a play or show to last about two hours and anything longer starts to feel as though it is dragging, whereas the same period of time would feel too short to count as a morning's work. If we unexpectedly see someone at the wrong time we might not even recognise them. Cultures develop shared ideas of appropriate timings; how long you should stay when invited into someone's home, even how long you should know a partner before you consider marrying them. Exceptions surprise us. I remember sitting at a lunch in Ghana on a table with six men, two of whom (one local and one from Scotland) surprised the rest of us with their tales of marriage proposals on first dates. (In case you are wondering, both the women they asked said 'Yes', and both marriages are still going strong more than two decades later.)

Routines give us a sense of security. They are so important that the mere act of breaking them can disrupt a person's concept of time and, in extreme cases, even cause terror. At Guantanamo Bay it was standard practice to make the timing of meals, sleep and interrogation unpredictable, defeating a prisoner's urge to count time and thereby inducing anxiety. Knowing the exact date was of no practical use to Alan Johnston, yet he knew he needed to try to keep track of the calendar. This desire for predictability and control is nothing new. In the early Middle Ages, Benedictine monks decided that predictability was essential to living a good and godly life and would ring bells at fixed intervals and carry out regular services to create a shared routine.

Time dictates the pattern of our lives – when we work, when we eat, even when we choose to celebrate. Just as the Benedictine monks knew when to expect the bells to ring, we each form appropriate temporal schemata for our own lives, which overwrite each other as they go out of date. (As soon as you have a new school timetable, it's very hard to remember the previous one.) Some of our temporal schemata are controlled by the changing patterns in the seasons, so inevitably winter and summer are particularly salient time-frames. Others are defined by our culture, so if I were to be dropped in my street at a random time and asked to guess the time, the day of the week and the month, then a combination of nature and culture would provide external cues to all three; if there is little traffic, few people walking by and no sign of life in the barber's, then it must be a Sunday. The temperature and the presence or absence of leaves on the sycamore trees will give a clue as to the time of year, and if the sun is out, its position in the sky would give a rough indication of the time of day.

The cyclical nature of the calendar helps us to organise time in our minds. When you are at school, the academic timetable punctuates the year, a punctuation that can have a lasting emotional impact (from which some teachers never escape). The American psychiatrist John Sharp has noticed that a number of his clients feel worse at the end of the summer – a hangover from years of back-to-school dread. Surprisingly, in the temperate climates in the northern hemisphere suicide rates are higher in the spring, as though deep despair sets in when the promise of spring fails to deliver respite from a spell of misery.

As you'd expect, the effect of the seasons varies depending on where you live, as do attitudes towards time. To investigate this the social psychologist Robert Levine compared the tempo of life in 31 countries around the world using three indicators. First, he measured the average walking speed of random pedestrians walking alone in the morning rush hour along a flat street with wide pavements. How fast did people choose to walk? Window-shoppers were excluded on the basis that they dawdle, and the streets selected were not so congested that the crowds would slow people down. Second, he wanted to compare the efficiency of an everyday task, so he measured the time taken to request a stamp in the local language, to pay for it and to receive the change. Finally, to establish the value placed on time-keeping in each culture, the accuracy of 15 clocks on the walls of banks in each city was checked. Combining these measures gave him an overall score for pace of life. It may not surprise you to learn that the USA, northern Europe and South East Asian countries had the fastest tempos, but Levine's findings weren't all so predictable. The efficiency of stamp-selling in Costa Rica brought the country up to thirteenth place in the tempo charts (funnily enough that's the exact opposite of the experience I had buying a stamp there, but then that's why we have systematic research on these things, rather than relying on anecdote). Even within the same country the variation can be extreme. On comparing 36 cities in the USA, in this instance combining walking speeds and clock accuracy with the time taken to obtain change in a bank, Boston came out fastest, while the home of showbiz, Los Angeles, was slowest, let down by particularly laid-back bank clerks. Everyone expected New York to come out on top, but in a 90-minute observation period during the early 1990s, the researcher witnessed one pedestrian dealing with a mugger and another with a pickpocket, which might have slowed them down.

At the time of the study the countries with the fastest tempos were also the countries with the strongest economies. This raises the question of which comes first – do people in active economies move faster because time is perceived to be more valuable, or did the fast pace of life lead to economic success? There's no doubt that energy and speed can help some businesses, but in some cases there is a limit to the extent to which the speed of your work can increase the market for your goods. However fast you make umbrellas, if it never rains where you live, no one will buy them. So the relationship between tempo and gross domestic product is best seen as a two-way interaction. Speed leads to some economic success, but economic success also requires people to move faster and makes a society more reliant on the clock.

TIME'S SURPRISES

So our minds create for us an experience of time which not only feels smooth on the whole, but which we can share with others, allowing us to co-ordinate our activities. Despite this, time never stops surprising us. The reason time is so fascinating is that we never appear to become accustomed to the way it seems to play tricks on us. Throughout life, we find it warps. We comment on weeks that seem to rush by, while others drag. We fly into a time zone that is behind us and create the illusion of cheating time, of living a few hours of life twice. Fly the other way, and we wonder what happened to the time we missed. Despite the longer evenings we get when the clocks go forward in spring, there is still a nagging feeling that an hour has been stolen from us. And when the clocks go back in the autumn we feel a sense of satisfaction at gaining an extra hour which marginally lengthens the weekend. The White Night festival in Brighton on the south coast of England, and its sister event Nuit Blanche in Amiens in France, were established to explore the theme of how you might use that extra hour in the middle of the night. You can do everything from listening to music in an aquarium to learning to knit in a bar. Although our rational side is well aware that this extra hour is just a trick of the clock, we still feel we are losing or gaining time, and this begins to illustrate how much of our relationship with time is based on illusions we create in our own minds.

In 1917 the wonderfully named researchers Boring and Boring conducted an experiment in which they woke up sleeping people and asked them to estimate the time, something the participants (including Mr and Mrs Boring themselves) were usually able to do successfully to within 15 minutes. But not everyone can do this. Although most of us find time slightly mystifying, for some of us it is utterly inscrutable. Eleanor is 17 and tells me she has never quite 'got time'. She is aware that she cannot judge its passing in the same way that everyone else seems to. When she wakes up in the morning, unlike the people in Boring and Boring's study, she has no idea what time it is and this continues all morning. She does not seem to sense time moving on. 'I don't know the time until lunchtime, when I start feeling hungry. I deliberately look for clues like that to work out how much time is passing.' At school she finds that while other people are able to make a rough guess, she can get the time wrong by several hours. Without checking the clock she has no idea whether a lesson is near the beginning or about to end. She inadvertently leaves her mother waiting where she has come to collect her because time doesn't feel as though it's passing, so she forgets to check her watch. So far the inconvenience has been mainly for her patient parents, but now that she's taking exams, she's beginning to notice the problems this lack of time perception can cause. While other students plan how much time to spend on each question, unless Eleanor constantly monitors the clock she doesn't notice that it might be time to move on. Her case illustrates that we don't all share the exactly the same concept of time. Eleanor also has dyslexia and this could hold the key to her difficulties with time perception. There is an intriguing link between the two, one which I'll return to when I discuss how the brain measures time.

For Eleanor time is constantly surprising, but in some circumstances it can be just as unnerving for the rest of us. We marvel, somewhat anxiously, at where the weekend went and how fast other people's children seem to grow up, or despair at how time drags in an airport queue. Imagine you are watching the final five minutes of a football match, and how differently that time passes depending on whether your team is winning or losing. If they're 1–0 down five minutes simply isn't long enough. If they're 1–0 up, time appears to stretch, giving the other team far more chances to level the score than they deserve. Think of a journey and how the way back always seems shorter. With fewer new memories to fill the time, everything seems familiar and it feels as though the distance is much shorter, unless, as the nineteenth-century philosopher and psychologist William James observed, you are retracing your steps because you have lost something. Then it seems endless. Time plays tricks on our minds.

As young children grow up, these mysteries of time are something they begin to observe for themselves. I asked two brothers what they had noticed about time passing. 'When you have to brush your teeth for two minutes that seems like a long time, but when you're watching TV two minutes goes really fast,' said eight-year-old Ethan. His 10-year-old brother Jake observed, 'If you're waiting for someone in the car while they go shopping it seems longer than if you do the shopping yourself.' These children have already noticed that time is deeply subjective. Our sense of time passing can even depend on the way we feel about our physical well-being. The psychologist John Bargh gave people anagrams to solve, then noted the time it took them to walk to the lift to go home after the experiment. Half the people were given anagrams of everyday words, but half were given words that might be associated with older people, such as 'grey' and 'bingo'. When these people walked to the lift, these subtle hints about old age had primed them to such a degree that it changed their sense of timing and they walked more slowly.[7]

So what are the major factors that cause time to warp? The first is emotion. An hour at the dentist feels very different from an hour working up to a deadline. If we look at pictures of serene faces we are quite good at guessing how long we watched them for, but show us a series of frightened faces and we overestimate the time that passed. However, the best illustration of the power of emotion to skew our perception of time is more dramatic – the slowing down of time when you are fighting for survival; when, like Chuck Berry falling through the sky, you are genuinely in fear for your life, one minute becomes elastic and can feel like fifteen.

TIME SLOWS DOWN WHEN YOU'RE AFRAID

Alan Johnston had long known that kidnapping was a risk that came with the job of a foreign journalist in

Gaza. It was an eventuality he had rehearsed in his mind before it happened in real life. When that fated day came around, and he saw a man get out of a car holding a pistol, his initial thought was, 'So this is how being kidnapped feels and this time I'm not just imagining it.' Then for a while everything went into slow motion. 'You can almost stand back and watch yourself going through it,' he told me.

Several weeks after he was captured, his captors gave him a radio. One night he heard a story on a BBC World Service news bulletin that made time slow down once more. 'They said that I'd been killed.' He began to think that perhaps the kidnappers' public relations department had got ahead of itself and released the news too early. Was this what they were planning to do tonight? 'It seemed more likely that they would want to keep me alive because that would be more useful to them, but when you're lying in the dark hearing that message going out to the world and they say they've killed you, there's a part of you that wonders if they're going to do it. Maybe tonight's the night.' For Alan, this felt like the longest night of his four months in captivity. Time definitely slowed down.

When people are afraid they might die, whether in a situation like Alan's, in a plummeting glider like Chuck Berry's, or in a car accident, they often report that the event lasted far longer than was possible. Somehow in just a few seconds they find the time to consider a great number of topics in detail. They think through their past, they speculate on the future and all the while they scan their memories for any piece of knowledge from anywhere which might help them to survive. This experience of time deceleration through fear is well-established, and – provided you feel frightened – time can distort even in a non-life-threatening situation. When people with spider phobias were instructed to look at spiders for 45 seconds (I am amazed they ever agreed to take part in this experiment), they overestimated the time that had passed. The same happened with novice skydivers. If they were watching other people, they gauged the duration of the fall to be short, but once it was their turn, time seemed to move more slowly and they overestimated the minutes they spent in the sky.

THROWING PEOPLE OFF BUILDINGS

Is this deceleration of time simply an illusion, or does the way we process time actually slow down when we are in fear for our lives? If the brain does process time differently when we are terrified, then it should also be able to process sights that are usually too rapid to see with the naked eye. To discover whether this is true, all you need to do is to scare people out their wits and then give them a test during that terror. One man knew just how to do it, and – in what appears to be something of a theme in research on time perception – was prepared, along with his brave volunteers, to go to extraordinary lengths to achieve it.

On the day of the study it was particularly windy. This was perfect. For the 23 volunteers standing on top of a tall tower in Texas, the wind injected a little extra anxiety into an already fraught situation. If this experiment was to work, real fear was essential. The neuroscientist David Eagleman, from Baylor College of Medicine in Houston (the same neuroscientist who wrote the best-selling imagined stories of the afterlife, *Sum*), warned his volunteers to stay well back from the edge until it was their turn to climb up inside a 33-foot-high metal cage mounted on the roof. He radioed down to the team on the ground 150 feet below to check that everything was ready, then turned to a line of digital wristwatches with giant faces. These perceptual chronometers were set to alternate very quickly between two screens showing random numbers. They flicked so fast that to the naked eye they looked like a blur. Eagleman wanted to know whether terror would speed up a volunteer's sensory processing enough for them to read the numbers which the calm human brain fails to register. Perhaps it's not that time slows down when we're afraid, but that our minds speed up.

Eagleman had previously experimented with taking the volunteers on a rollercoaster, but they just weren't scared enough; and in fact many seemed to enjoy the experience. It was time for something more drastic – freefall. Eagleman knew that no one would agree to take part in this experiment unless he had shown that he was willing to do it himself. Strapped into a harness, he was dangled over the side of the tower block and dropped, backwards. (Forwards wasn't sufficiently frightening.) Then he did it again. And then again. Before the third attempt he was convinced that he would be less terrified; experience would surely tell his brain that he would be fine. But no, he told me, 'It was still

beyond scary.' Then it was the turn of a young man called
Jesse Kallus. Just as Eagleman had been before him, Jesse
was thrown off the building and by the time he had been
caught safely at the bottom he had reached a top speed of
70 miles an hour.

Everyone who took part in the experiment reported that
time felt as though it decelerated. The fall stretched every
one of those unbearably petrifying seconds. So the first
element of the study had worked; the desired effect of
subjective time dilation had been achieved. Yet still the
figures on the watch face flickered too fast for their brains
to perceive them. David Eagleman had demonstrated that
time itself doesn't actually slow down when we're afraid,
and nor does the brain's sensory processing speed up. What
changes is our perception of time – our mind time.

So how does this happen? It is true that fright does etch
strong memories into the brain, and – as will emerge in
this book – memory is one of the key factors in making
time warp. When people are shown a video of a bank
robbery lasting exactly 30 seconds, two days later they tend
to guess that it lasted five times longer than it did. The
more disturbing a version of the video they are shown, the
greater their overestimation of its duration.[8] After a stressful
event we often recall every single detail of what we saw,
heard or even smelt. The richness and freshness of these
memories contributes to our sense of how long it lasted.
We become accustomed to a certain quantity of memories
fitting into a certain time-frame. Usually this serves us well,
but during a life-threatening incident the intensity of the
experience results in the creation of more memories. Every

second feels brand new, which causes us to judge the event to have taken longer than it really did, to have happened in slow-motion. This sensation is amplified by the fact that in a car accident, for example, the mind focuses on the elements of a situation necessary for survival and filters out anything inessential such as the scenery, the songs changing on the radio or the number of cars that pass. These are the cues which would normally help to assess time passing. Without them, once again time warps.

The big question is whether the combination of the plethora of memories and the absence of cues to time passing is enough to make time decelerate this drastically? There is a more radical explanation – is it possible that the way the brain actually measures time could make it feel as though it slows down? If the brain counts time by monitoring its own processes, when it moves extra fast in an emergency this could cause it to count more beats and to believe that more time has passed. So while the brain is racing to save itself, so is its clock. I'll come back to this in the next chapter. Before that there are other curious factors that distort time. The life-threatening, mind-racing moments of intense concentration are not the only occasions when time decelerates. The opposite – having *nothing* on which to fix your mind – in other words sheer boredom – has a similar, though less extreme effect, as do a series of other experiences.

NOT THE KINDEST OF EXPERIMENTS

You arrive to take part in a study. You know it is to take place in the psychology department, but not what it involves.

There are five other participants, all wearing name-tags. Everyone seems friendly, if a little unsure of what might be about to happen. The woman in charge says that first you should get to know each other and she gives you a list of topics to discuss, among them the place you would most like to visit in the world, your most embarrassing experience and what you would choose if you could have one magic wish. Soon you're happily exchanging tales of humiliating incidents, like the time you got an electric hot-brush stuck in your hair on the way to a wedding and had to walk along the street with the flex hanging down from your head (this happened to me). The psychologist says you will be working in pairs and, to make things go smoothly, you should write down the names of the two people from the group with whom you would prefer to work. That's easy. You hand in your form and wait to see who you'll be paired with. But when they call you in for your turn they look rather embarrassed and say that no one has put you down as someone they'd like to work with. They say that in all the studies they've been running, this is so unusual that they think it best for you to work on the tasks alone. You're a bit surprised, and – if you're honest – hurt, but you try to tell yourself that it doesn't really matter what a group of strangers thinks of you and that you didn't particularly like them anyway. You're determined not to show anyone that you're upset and to do the tasks as well as you can. For the first task, they start a stopwatch, then stop it and ask you to guess how much time has passed.

While you sit alone wondering why no one likes you, what you don't realise is that every other member of the group has also been taken aside to separate rooms to work alone, but while half were given the same explanation they gave you, others were told they must work on they own because they had been chosen by everyone, making it difficult to allocate partners fairly. A harsh experiment you might think, although not as bad as a study later in the same series where they tell you that the results of your personality questionnaire indicate that although you might marry several times, none of your relationships will last and you are likely to spend your old age alone. I should add at this point that after all experiments like this participants are debriefed and told it's all a fiction.

The intriguing result of this rejection study is that the belief that a few strangers dislike you can alter your time perception. The people who were told they were popular estimated the 40 second test as lasting an average of 42.5 seconds, while the rejected group came out with an average of 63.6 seconds.[9] Although 20 seconds might not sound like much, the fact that there was a difference is quite extraordinary. The rejection had made them painfully aware of everything happening in the present. Their misery had stretched time.

This research on rejection and time perception stemmed from the work of psychologist Roy Baumeister, who studied people who were contemplating suicide. Those in this situation tend to experience what is known as a deconstructed state, where they have such a strong sense of an inner numbness that they have little or no concept of a future and find it hard to imagine that life might ever improve if they remain alive or that choosing death would

have serious ramifications. People planning suicide are in
a very particular mental state where the perception of time
can become skewed. As an aside, it is a state which can
also explain why suicide notes often reveal so little. The
American sociologist Edwin Shneidman spent more than
a quarter of a century studying the meaning of suicide
notes after finding a collection of them in the vaults of the
Los Angeles County Coroner's office in 1959. He decided
to devote his career to their study, determined to gain an
insight into the suicidal mind. His analysis showed, probably not surprisingly, that suicide notes contain a greater
percentage of first-person singular pronouns than other
sorts of documents. But it seems in terms of insight their
content is disappointing. After spending more than 25 years
obtaining and analysing notes, Shneidman concluded that
most tell the same story, and despite being 'written at
perhaps the most dramatic moment of a person's life, are
surprisingly commonplace, banal, even sometimes poign-
antly pedestrian and dull'.[10] Later in life he decided that
odd phrases could sometimes be telling, but that most notes
still bring little by way of explanation to those left behind.
Only a third of people who kill themselves even leave a
note. Somewhat harshly Shneidman believes that those who
do are the kind of people who like stating the obvious. He
doesn't disguise his bitterness at his disappointment with
the style of the notes: 'To a "Quarantine – Measles" sign
such a person might add the words "Illness inside – please
stay out".' He believes that because people who are about
to kill themselves are in this altered state, a state of such
fixed purpose where time is warped, they are unable to

explain much about their state of mind. The tragedy here
is that an explanation is exactly what those left behind are
searching for. And Shneidman believes we optimistically
look for even more than that; we hope that someone on
the brink of death might have some 'special message for
the rest of us'. But lest we think that Shneidman was lacking
in sympathy for those driven to suicide, he did a great deal
to pioneer the field of suicide prevention and co-founded
the Los Angeles Suicide Prevention Center in 1958, a centre
that was to become famous in 1962 after it concluded that
Marilyn Monroe's death was caused by 'probable suicide'.

People with depression can experience distortions of time
even if they are not feeling suicidal. During an episode of
depression, the past and the present become central, while
the future – especially any kind of hopeful future – is almost
impossible to imagine. The British psychiatrist Matthew
Broome has frequently seen this in patients. And experiments confirm that people with depression give time estimations that are on average twice as long as those who are
not depressed. In other words time is going at half its
normal speed. This leads me to wonder whether in some
cases depression could be considered a disorder of time
perception. Or perhaps the slowing of time is a consequence
of depression, which then helps to maintain it and makes
it harder to escape from. Matthew Broome points out that
we know that sleep deprivation and the use of a light box
can both elevate a person's mood as they confuse the
internal clock.[11] When a person is depressed the present
and the future become 'bound to one another in suffering'.[12]
The effect is so distinct that the philosopher of psychiatry

Martin Wyllie suggests that as an additional diagnostic tool, mental health professionals could ask their clients to estimate the duration of the consultation. I wonder whether you could simply ask them to estimate the passing of a minute. If 40 seconds feel like a minute to them, then time is stretching. The more slowly time is passing for that person, the more severe their condition might be.

Time also decelerates for the most anxious cancer patients. The psychophysicist Marc Wittman has found that they overestimate time intervals and report that time seems to be slowing down. Contemplation of their mortality has directed their attention to the passage of time with the result that it protracts it.[13] In contrast, for patients experiencing conditions involving a break with reality, such as schizophrenia, time can distort in many different ways – appearing to vary in speed, repeat itself or even stop altogether. The Cotard delusion takes this distortion of time perception to extremes. Named after the French neurologist who first described it in 1882, the Cotard delusion is a rare condition of extreme pessimism, beginning with depression and ending in the denial of everything, including possession of the main organs of the body, having a family, a future or even an existence. Back in 1882 Jules Cotard wrote of one of his patients, 'Stating that she was no longer anything, the patient begged for her veins to be opened up, so that it could be seen that she had no more blood and that her organs no longer existed.'[14] In a sense this is the ultimate disorder of time. There is no sense of a past or a future, and three-quarters of the patients in the subsequent case reports of this rare condition even believed that they were dead.[15] It is very rare, but, as we'll see, problems with the perception of time could also be a root of a far more common condition.

HYPERACTIVE TIME

He doesn't sit still. He fidgets. He can't concentrate. He moves impatiently from one thing to the next, constantly getting distracted. This might sound like the description of any lively child. But there is a big difference. Children with attention deficit hyperactivity disorder, or ADHD, do these things far more than other children of the same age and it has been discovered very recently that faulty timing might be the key. Children with ADHD are rooted in the present. They find it hard to consider the consequences of their actions and they find waiting, even for a short time, excruciating. This might be because what feels like five minutes to the rest of us, feels like an hour to them, so when they are told to sit and wait for five minutes this could be a task they find genuinely challenging. In laboratory experiments children with ADHD find timing tasks very difficult. Their experience of time appears to be different from that of other children. If they are asked to say when three seconds have passed they think they're over in far less than that; in other words if you have ADHD time passes very slowly. This finding is so common in children with ADHD that Katya Rubia, a cognitive neuro-scientist at the Institute of Psychiatry in London, has been able to use time estimation tasks to correctly classify 70 per cent of cases, quite a feat considering there are currently no conclusive tests for ADHD; current diagnoses rely on experts watching a child's behaviour and then making a judgement.

It seems remarkable that the most common childhood disorder, affecting between 3 and 5 per cent of all children, could be down to timing. It manifests itself in various ways. If I were to ask you whether you'd like £100 now or £200 in a month's time, most of you would go for doubling your money, but for people with ADHD, delayed gratification is unappealing. If children with ADHD are asked to watch for a red light to come on, wait five seconds and then press the button in order to get a prize, they are so keen to press the button that they can't resist pressing it straight away. Children with ADHD find it very hard to wait and often act prematurely, without considering the consequences. While many of us strive to live more in the present, these children live too much in the present.

If ADHD is a disorder of time perception, could you somehow change a child's relationship with time and in turn reduce the symptoms of ADHD? At the moment therapeutic intervention tends to focus on inhibition and helping children to think before they act, but Katya Rubia plans to develop a form of cognitive behavioural therapy where children are taught how to wait and how to delay. This is something I'll come back to in Chapter Five. The difficulty is this: if a child experiences the passage of time in an unusual way, teaching them to wait won't eliminate the fundamental problem. They might learn to tolerate the aching slowness of time, but if a five-minute delay feels like an hour, then it always will. They might be able to learn not to behave impatiently, but to them wouldn't it still feel like an agony of time? Here Katya is optimistic that the brain's plasticity is such that if she can teach them to behave differently, then this could eventually have an impact on the brain and on time perception itself. She has already demonstrated that Ritalin, the drug commonly used to treat the symptoms of ADHD, does improve time perception and the estimation of milliseconds. Perhaps learning to wait would give children the opportunity to learn to judge a time interval more accurately. As Katya told me, 'If you never wait, you probably don't learn to estimate a time interval properly.'

To sum up: so far it is clear that ADHD, extreme fear, rejection, boredom and depression can all lead to the sensation that time is slowing down. The next situation which can dilate time is altogether more surprising.

DIVING FOR TIME

There were fourteen scuba divers in all – six amateurs and eight Royal Engineers. It was a hot August day in Famagusta Bay in Cyprus in the mid-1960s. The resort was fast becoming the place to be seen. New hotels were appearing; ready to accommodate the rich and famous on holiday. Archaeological excavations in the long arc of sand were slowly revealing a perfect oblong of pillars outlining the site where an old gymnasium stood, until, according to legend, in the fourth century BC the king burned down his Palace of Salamis rather than submit to the Egyptians.

But the 14 scuba divers were not here to admire the archaeological sites, nor even the grouper fish and Spanish lobsters under the water. They were here to take part in a study on time. At the start of the experiment, each diver sat with a thermometer in his mouth while his pulse was taken. Then, without counting, he had to guess when a minute had elapsed. Next a Royal Engineer handed him a one-ounce charge of gun cotton and lit the fuse. The diver's job was to take the fuse, swim down 15 feet to place it on one of the many shipwrecks submerged beneath the waters of Famagusta Bay and then return to the surface to wait for the explosion. Then the initial routine of sitting on the deck while his pulse and temperature were taken and estimating the passing of a minute was repeated. But here was the catch. The divers were instructed that if the charge did not explode within a few minutes they were to dive back down to the shipwreck to retrieve the gun cotton. These explosions were genuine, so not surprisingly this injected an element of anxiety into the experiment. It was conducted by Alan Baddeley, who was later to become one of Britain's most eminent researchers in the field of memory. He was in Cyprus to follow up an experiment he had conducted one March day in the cold waters off the coast of Wales. He had discovered (no surprises here) that the divers were colder after their dive, and that the colder they were, the longer they estimated one minute to be. In other words for them time felt as though it were going fast (if this sounds strange to you, remember that if time had felt slow, they would have *under*estimated the minute passing, feeling that after 40 seconds it must surely be over). However it was possible that instead of time speeding up *after* the dive, their anxiety might have slowed time down *before* the dive and that this might be the explanation for the discrepancy in the before and after timings. So he relocated his experiment to the warm waters off Cyprus and devised a task where the divers' body temperatures would barely change, but which was extra stressful, due to the inclusion of the explosions. In the experiment in Cyprus there was hardly any difference in the speed of their counting, before and after the dive, supporting his original idea that it was temperature that was changing the perception of time in the Welsh divers, not anxiety.[16]

Three decades earlier the wife of an American psychologist called Hudson Hoagland was lying in bed with flu. Although her husband was caring for her kindly, she complained that whenever she needed him he seemed to be absent from the room for long periods. In reality he was only away from her for a few minutes at a time. Wondering whether her experience of time was askew, he took the opportunity to conduct an experiment on time perception and body temperature. Her fever was causing extreme fluctuations in her body temperature so every time the thermometer gave a new reading, he asked her to count the seconds passing until she reached one minute, all the while monitoring her accuracy with a stopwatch. And just to be on the safe side, at each temperature he persuaded her to perform the counting task five more times, meaning that in the space of 48 hours his ailing wife took part in 30 trials for the experiment.[17]

He discovered that not only was she a very patient patient, agreeing to his constant requests for her to spend a minute counting without knowing why, but that the higher her temperature, the sooner she thought a minute had passed. When her temperature reached 103 degrees, time had slowed to the extent that she thought a whole minute had passed after just 34 seconds.

Hoagland must have possessed strong powers of persuasion because for his next experiment he convinced a student to submit to diathermy – that is for his body to be wrapped up tightly and then artificially raised to 38.8 degrees using an electric current. Bearing in mind that a body temperature of 40 degrees would be considered a potentially life-threatening emergency, the student was unsurprisingly rather anxious, which Hoagland remarked rendered his initial time estimations somewhat erratic. Once the student had managed to relax, his perceptions of time were altered in the same way they were for Hoagland's wife. As his temperature rose, time decelerated. Hoagland tested just two people, but Baddeley's later work with the divers confirmed that body temperature can warp our experience of time.

FIVE TIMES A DAY FOR 45 YEARS

The discovery of the next factor that can slow down time required great dedication, something this field of study does seem to engender. Robert B. Sothern is a biologist who has been taking a series of measurements every single day since 1967. Five times a day he estimates the passage of a minute without looking at a clock; measures his blood pressure, body temperature and heart rate; tests his eye-hand co-ordination and rates his mood and vigour. For 19 years he even co-opted his parents to help with the task and for several decades he also recorded data on the strength of his grip and the volume of his urine. It all began after he volunteered to travel from the United States to Germany to take part in an experiment where he lived underground for three weeks without any means of keeping time. This experience gave him the idea of investigating how his rhythms changed as he aged, using that most willing of participants – himself. Where else could you find a subject so motivated and conscientious that they let neither holiday nor illness disrupt the research process? Robert has now conducted more than 72,000 measurement sessions and tells me he has no plans to stop.

Robert's main interest is in how the timing of medical treatment might affect its efficacy. Does it work better in the morning or the evening or on a particular day of the month? It's a field that he acknowledges is regarded with scepticism by the medical community and, seeing the sparsity of the evidence, it's likely to remain so. But what interests me is a sideline of this research. His decades of measurements of time-estimation reveal another factor which slows down time – youth. During his period of isolation in Germany his time estimations showed that for him time was decelerating. But as he left his twenties the opposite happened and time appeared to be gradually speeding up.[18] This is a common sensation as people get older, and one that I'll explore later in the book.

HOW TO MAKE TIME STAND STILL

So emotions, fear, age, isolation, body temperature and rejection can all affect our perception of the speed of time, as does concentration, or 'attention' as it tends to be referred to in the psychological literature. If you happen to be in a room that has a clock with a second hand that ticks rather than sweeping round smoothly in one motion, glance up at the clock face and see what happens. If by chance you catch it at the right moment the second hand will appear stationary for longer than it should. You wonder whether the clock has stopped, only for it to start moving again a moment later. This is a demonstration of chronostasis: the illusion that time stands still. If it doesn't work the first time, glance up a few more times and eventually it will. The traditional explanation for this illusion is that in order to present us with a consistent image of the world that doesn't blur every time we shift our gaze across the room, our brains momentarily suppress our vision whenever we move our eyes. The result gives us the impression of life as a smooth film. In order to compensate for this moment of suppressed vision we assume, not unreasonably, that most objects in a room are stationary. The ticking second hand tricks our brains. Or that's the theory. The problem with this explanation is that the clock illusion occurs with other senses too. A similar phenomenon known as the dead phone illusion happens in countries where the dialling tone consists of beeps interspersed with silence. If you pick the phone up at the right moment the initial silence feels so long that you get the impression that the phone is dead.

So what does this have to do with attention and the warping of time? Well, the researcher Amelia Hunt has an alternative explanation for the clock illusion, one that sheds light on the way attention can affect time perception. We can catch a ball or drive a car safely while constantly gauging times with precision, but overt timings are more difficult to get right.[19] Her explanation for the clock illusion has nothing to do with vision and everything to do with attention. Time, she suggests, is distorted because we have glanced across the room and are concentrating on something new. When we focus our attention on an event, even one as brief as looking at the clock, it creates the impression that it lasted longer than it did. Attention can also explain why boredom slows down time. Writing in the nineteenth century, the influential psychologist and philosopher William James suggested that boredom occurs when 'we grow attentive to the passage of time itself'. To illustrate this sensation, he suggested closing your eyes and getting a helpful person to tell you when a minute has passed. Try it: it seems like ages. And that silent minute will seem even longer if the preceding minute was filled with music or speech. Likewise the involvement of attention can explain why rejection slows time down. The rejection causes us to focus in on ourselves and our shortcomings, and once again time is stretched.

Whether we're falling through the sky or watching a clock, it is becoming clear that our relationship with time is not straightforward. Attention is just one part of the story; our shared understanding of time is another; and in the next chapter I'll ask how it is that the brain measures time at all, when there is no specialised sense organ for time.

Meanwhile we left Chuck Berry suspended in time in

mid-air on the New Year's Day gliding trip that had gone
so wrong. By now he would surely have crashed to ground.
Standing on Coronet Peak, his aviator friends had heard a
bang. They watched as the wings fell off the glider and saw
that Chuck was starting to fall, seemingly dragging the
remains of the aircraft behind him. Then he disappeared.
Why wasn't he opening the reserve chute? Without it there
was no way he could survive.

With so much to think about Chuck had not felt particularly frightened; even though time had expanded, he didn't
have time to be scared. He stretched his arm as far as he
could and finally found what he had been searching for,
the handle of the chute, flapping in the wind. He yanked
it hard, yearning for that comforting sensation that comes
when the canopy bursts into bloom and you begin to rock
gently in your harness, as though picked up and cradled
by a giant. But that didn't quite happen. He began to slow
a bit, but knew he was still falling too fast. Looking up, he
understood why. The chute was old-fashioned, small and
round. 'Like the ones the airmen had on D-day?' I asked
him. 'Like that, but ten times smaller.' Now he was scared.
After all that he'd gone through, he was going to crash
anyway. If only there were some trees. Usually he would
do anything he could to avoid landing in trees, but at this
speed and with a 2,000-foot drop, a fall broken by branches
could be his only chance of survival. But there were no
trees nearby, just the bushes on the steep slope of Coronet
Peak. Time had been passing achingly slowly. Now everything changed. It was fast. There was no way of steering,
and he crashed into the bushes.

Half an hour later he was still lying on the ground, strapped into the wreckage of the cockpit. He had no idea how he had got there. Looking down at his clothes he realised he must have been gliding, but here he was stranded on a hillside without a glider. Then he saw the Swift's wings higher up the hill.

The global positioning satellite system in Chuck's pocket provides some unintended data on time perception. It too survived the crash. So while Chuck's perceptions of the accident might tell him one story, the GPS and its accurate records of his precise location at each moment in time tells a different one. 'That freefall took forever. It was the longest time.' In fact that everlasting fall had taken just 10 seconds and the hurtle to the ground with the tiny parachute had lasted another five. After his crash, Chuck remembers calling the air traffic control tower in Queenstown to inform them about the accident. He only remembers one call, but the phone shows he spoke to them twice, suggesting he was confused, if not concussed. He lay high up on the side of the hill waiting for his rescuers. It was 40 minutes before they reached him, but now time was playing tricks on his mind again. It was speeding up. He was so elated that he was convinced they arrived in 10 minutes. 'I was just stoked to be alive, really. There's nothing better.' And as for injuries, he told me, 'I had a bump on the head and a prickle in the wrist. That was it.'

He puts his survival down to his many years of experience skydiving. To him freefalling feels normal, so he didn't panic. He's not given up on the adventure sports either and is now building his own plane. Chuck believes that two decades of skydiving have changed his perception of time, and not just when things are going wrong. To most of us five seconds seems like a short time, but he knows that it's long enough to travel 1,000 feet when you're falling. He now thinks that five seconds is a long time. His experience is a good illustration of the way we each create a sense of time in our minds. To understand how we do it, it is necessary to look at the way the brain counts time.

MIND CLOCKS

THE ALARM WENT off at 5.00 a.m. It was the wet season in Costa Rica, and we had got used to downpours in which it seemed as though huge vats of water were being tipped from the sky. This morning though was calm and dry: ideal for a bird-watching expedition.

Ricky arrived exactly on time and spent some minutes carefully making a piratical skull cap out of his blue bandana. Once that was in place he topped it with a brown cap with the peak facing backwards. This headgear was decorated with an arrangement of twigs, leaves and bird feathers. Greying dreadlocks spilled out from under his hat and his weathered features were grizzled with a straggly, gorse-like beard. Our guide was a winning combination: Rastaman meets British naturalist David Bellamy.

Ricky, like many Costa Ricans, is a mixture of ethnicities. In his case part Afro-Caribbean and part Bribri, one of the indigenous groups of the country. As a boy he followed his friends in trapping birds and keeping them in cages. But unlike his friends he was never cruel. His grandmother taught him to care for the birds – to admire them and then release them. As an adult he became a naturalist, and now he takes people from all over the world to see birds on Costa Rica's Caribbean coast.

It was grey dawn and the sun couldn't break through the cloud cover. This made the colours of the birds hard to see, but the Samasati valley was full of their sounds. Soon we heard the grating call of toucans and saw two flying overhead. When they landed at the top of a tall tree in the distance, one glimpse through the binoculars made it easy to see why so many tour agencies choose toucans for their logo. These were keel-billed toucans with green, red and yellow striped beaks with a streak of lime green across the top. In another tree we spotted a black-cheeked woodpecker that mimicked the exact movement of a toy I had on a spring on the end of my pencil at junior school.

And it wasn't just birds that we saw. A bulky grey blob wedged in the fork of a tall, leafless tree turned out to be a female two-toed sloth. Sleeping, of course. Ricky told us that she would stay up there for days, venturing down only for her weekly defecation. Sloths are rather fastidious about their toilet, burying their deposits in the way cats do. This devotion to hygiene comes at a price, however, as many end up being killed by dogs while going about their ground-level business.

The morning was beginning to heat up and take on the familiar sticky humidity. We were tiring slightly. Then Ricky saw it. This was the bird we'd come to see, a bird with some most unusual skills – the rufous-tailed hummingbird. It was so small that it could almost have been a flying insect.

Weighing less than a large paperclip it hovered in mid-air, dipping its curved red beak into the flower heads, its wings whirring in a figure of eight too fast for the human eye to see. What we could see was its emerald green head and the famous rust-coloured tail.

Hummingbirds, or 'hummers' as their biggest fans like to call them, are the only birds in the world that can fly backwards. Quite a trick. But what is also fascinating about the hummingbird is its ability to judge the passage of time. Just as humans can guess when 20 minutes have passed, so can hummingbirds.

They visit a plant, hover there, wings ablur, while they dip their stick-thin bills and elongated tongues into the long flower tubes and suck out the nectar. Having had their fill, they move on. The rufous-tailed hummingbird protects its source of food by aggressively seeing off any other birds that enter its territory, but it has a second technique of ensuring it gets to the nectar before anything else does. This is known as trap-lining and allows the hummingbird to calculate exactly when 20 minutes have passed – the time it takes for the flower to replenish its nectar. By returning with such precise timing the hummingbird beats other birds to this life-giving substance.

So we know that hummingbirds can judge the passage of 20 minutes, but have they evolved to measure *only* this interval or could they somehow learn to judge shorter time intervals too? To find out, researchers at Edinburgh University created fake flowers with a nectar replenishment cycle of 10 minutes instead of 20. Could the hummingbirds in the lab learn to judge when 10 minutes had passed?

It turns out they could.[20] And it isn't just exotic birds that possess this remarkable skill. The everyday feral pigeon can be trained to judge time intervals with a fair degree of accuracy too.

As we saw in the last chapter, humans have this ability too. We can detect the millionths of seconds necessary to locate the direction of a noise, but we can also make a stab at guessing the year of every individual memory we hold. In this chapter, I will consider the competing explanations for how the brain copes with this range of time frames. It feels as though there must be a clock in the brain that ticks away the milliseconds, seconds, minutes and hours, allowing us to make judgements about time, but so far neither through dissection or ever-improving brain scanning techniques has a single clock structure been found. Just as Einstein's theory of relativity tells us that there is no such thing as absolute time, neither is there an absolute mechanism for measuring time in the brain.

We do have a body clock, but this only controls our 24-hour circadian rhythms. It has no role in the judgement of seconds, minutes or hours. What neuroscientists in this field are all trying to establish is how the brain counts time when there is no organ to do so.

Just as Chuck Berry's experience of time was dilated by terror during his long fall through the sky, and Mrs Hoagland's by her fever, it is clear that however the brain counts time, it has a system that is very flexible. It takes account of all the factors I discussed in the last chapter – emotions, absorption, expectations, the demands of a task and even the temperature. The precise sense we are using also makes a difference; an auditory event appears longer than a visual one. Yet somehow the experience of time created by the mind feels very real, so real that we feel we know what to expect from it, and are perpetually surprised whenever it confuses us by warping.

You can easily test your own skills at time estimation by starting the stopwatch on your phone, looking away and then trying to guess when a minute has passed without counting in any way. Most of us are fairly good at it, but there is individual variation and our skills decrease with age. We are also easily distracted; people can estimate the length of a song fairly accurately if that's all they are concentrating on, but if you ask them to focus on the pitch of the song as well, they will overestimate its duration. Not surprisingly, people who are particularly prone to boredom tend to give an underestimate of that minute passing. Time has dragged so slowly that they might think the minute is over in just 30 or 40 seconds.

Discussion of any of these studies can be confusing because there are two ways of measuring time estimation: prospectively – where you ask sometime to estimate a minute starting from now – or retrospectively, where you give them a task and then afterwards you ask them to guess how much time has elapsed. If time is moving slowly a person will *under*estimate the passing of a minute at the time, but if they're asked afterwards they will *over*estimate the duration. Both signify time passing slowly. Imagine you're at a play that is particularly dull. If while you sat impatiently hoping for the interval, you were asked to say when an hour had passed, time would be dragging so much that you might guess an hour had passed after only 40 minutes. When the interval finally does arrive you look back and insist the first half felt like two hours rather than an hour. So, glancing at the figures, one looks like an underestimate and one an overestimate, but they both indicate the perception that time is slow-moving.

Although no single clock of the brain has been discovered, several areas have been found to be implicated in time perception, each of which also reveal something about our experience of time. Let's begin with the cerebellum. This area at the back of the brain, down towards the nape of the neck, accounts for just 10 per cent of the brain's volume yet contains half of all our brain cells. The cerebellum, which means 'little brain', helps us to co-ordinate movement by processing huge quantities of information from the rest of the nervous system. It is thanks to this part of the brain that when we wake up in the morning we can immediately detect the position in which we are lying (a sense known as proprioception) because the cerebellum is constantly monitoring the position of each limb. Though this might sound inconsequential, having met Ian Waterman, who contracted a rare neurological illness at the age of 19 which severed the pathways sending messages from his body to his cerebellum, it is clear that this sense is vital. He has now learnt to learn to walk again and can drive a car, but in order to do so he must watch his own arms and legs continuously, consciously observing and thinking about every movement he makes. If he loses his focus for just a second, an action as simple as holding an egg results in the egg either smashed on the floor or crushed in his hand.

Ian's difficulties are caused by the loss of all sensation below the neck, which means that his peripheral nerves are unable to provide feedback to the cerebellum. With its sheltered position at the back of the brain it is rare for the cerebellum itself to become injured, but if it does it's not only the smooth co-ordination of movement that is disrupted, but the perception of the tiniest fractions of time.

If you puff a tiny amount of air onto someone's eyeball they will blink in discomfort, but if a signal is given beforehand, then, just as Pavlov's dogs began salivating at the sound of a bell, a person will blink at exactly the right moment in anticipation of the puff of air. Unlike Pavlov's classically conditioned reflex to salivate, this blinking demands precision-timing and it is the cerebellum which makes the calculation. The finding that a person with a damaged cerebellum loses this ability is so robust that in 2009 a team working with patients in Cambridge and Buenos Aires found that this air-puff test could be used to predict which patients in a persistent vegetative state might one day recover consciousness. But the strongest evidence of all for the involvement of the cerebellum in time perception comes from a more dramatic technique.

ELECTRIFYING THE BRAIN

When I was shown into the consulting room, an old lady was sitting on a chair in the middle of the room. She looked anxious. The doctor approached her head holding what looked like a giant version of those bubble blowers which children play with at birthday parties. It was attached by a long curly wire to a trolley packed with electrical equipment. I'm afraid when the doctor insisted, in a mittel-European accent, that 'zis is perfectly armless' I couldn't help but think I was in a sci-fi film, with the mad professor intent, despite his assurances, on electrocuting his elderly patient.

'Look at zis!' he said, placing the coil against his own head and flicking a switch. Suddenly one side of his top lip was twitching up and down in a sneer. 'And I can do zis.' He moved the coil to a different part of his head, switched the machine on again and one of his arms shot up in the air, in a slightly limp version of a Nazi salute. 'Do you want a go?' he said, lunging towards me with the big coil. I was sure I didn't.

The doctor was demonstrating equipment that induces convulsions through a gentler version of electro-convulsive therapy. The elderly lady was about to try an even milder variant. This weaker coil would simply stimulate a specific area of the brain through a process called Transcranial Magnetic Stimulation, or TMS. She was putting herself through the process because she was so depressed that she felt suicidal. Nothing else had so far made her feel better.

The doctor spent a long time examining her skull. When he was certain he had found exactly the right place he picked up the second coil, counted down from 10 and then applied a series of pulses to her brain. She moaned quietly, more in fear than in pain. But even so, she was hoping for some relief. In trials many people have found this treatment reduces their depression. She would now wait to see whether it would work for her.

The ability of this equipment to target precise areas of the brain makes it useful not only therapeutically, but also for identifying the parts of the brain involved in time perception. The electric pulses can temporarily disable a specific brain region without lasting side-effects and this has provided the strongest evidence to date regarding the involvement of the cerebellum in time perception. When this part of the brain was dampened down using TMS, people found it harder to estimate time. More specifically, it reduces people's ability to perform in tests where they have to judge milliseconds, but makes no difference when the time intervals are many seconds long. To assess those we need to use another area of the brain.

THE MAN WHO THOUGHT THE WORKING DAY HAD FINISHED

A man sat in the consulting room at the Santa Lucia Foundation on the outskirts of Rome waiting to see Dr Giacomo Koch. Back in the sixties the hospital looked after war veterans, now it specialised in neurological damage and the man was hoping the specialists would be able to help him. He was only 49 years old, yet was finding it hard to concentrate and for a few days he had felt a bit weak down one side of his body.

The case was an intriguing one. The man was convinced something quite serious was wrong, but the doctors could not diagnose a recognisable condition. They ran test after

test. To check his memory there was the Digit Span, the
Corsi Span, the Rey-Osterrieth Complex Figure test (immediate and delayed recall), the Verbal Supraspan and the
Forward and Backward test. To test his visuospatial skills
there were the Raven Progressive Matrices. For concentration there was the Trial Making test, for language the Verbal
Fluency and Phrase Construction tests. For decision-making
there were the Tower of London and Wisconsin Card
Sorting tests. The doctors calculated his scores. They were
all normal. They had him copy drawings, learn lists of words
and complete well-known phrases. Still his scores were
normal.[21]

But the man had also reported another strange sensation;
his mind clock and the actual clock seemed to be radically
out of synch. He would go into the office, do what felt like
a day's work, get ready to leave and find it wasn't even
lunchtime. On other occasions, events seemed to last for a
much shorter time than was the actual case – a minute to
him seemed like just 30 seconds.

In the light of this, the doctors performed some time-
estimation tasks. In order to get a fair baseline measure,
they arranged for eight volunteers, also in their forties,
to take the same tests. Each sat alone in front of a computer
screen while a series of random numbers appeared one
at time. Their task was simply to read the numbers out
loud. This would prevent them from counting time in
their heads. Afterwards all they had to do was to guess
how long the tests had taken. The results of one trial
could be down to chance, so they repeated the task with
different numbers until they had done it 20 times. In each

instance, the man performed worse than the eight volunteers. His ability to judge the passage of time was somehow impaired.

A brain scan indicated some damage to the right frontal lobe, which as the name implies is near the front of the brain on the right. This gives us a clue as to the next area of the brain implicated in time perception, an area we would normally associate with the ability to hold something in mind known as working memory. It is this skill that allows you to read through a recipe, then remember the list of ingredients while you go to the cupboard to fetch them. The very front of the frontal lobe, the pre-frontal cortex, which is located behind the forehead, seems to be particularly crucial.

The involvement of this area of the brain in time-keeping is backed up by the curious new finding that children with Tourette's syndrome have recently been found to be better than other children at judging durations of just over a second.[22] Suppressing their tics involves activity in the prefrontal cortex and experts have found those children with Tourette's who were particularly good at suppressing their tics did even better on the timing tasks. This suggests that their need to use this region of the brain to control their tics brings an added advantage in terms of time perception.

So far we have looked at two areas of the brain associated with time perception – the cerebellum, low down at the back of the brain for those millisecond judgements, and the frontal lobe, behind the forehead, for durations of seconds. But what happens when we try to judge much

longer durations of hours or even days without access to a
clock or any clues as to day and night?

THE PERFECT SLEEP

Do glaciers carve their way through caves underground in
the same way as they gouge through mountains above the
ground? This was the question that the French speleologist
Michel Siffre set out to answer when he planned an underground expedition in 1962. But having made his initial
arrangements, he began to ponder a quite different question,
one which was to revolutionise another field of study
entirely.

He would make the trip as planned, taking all the standard
equipment such as tents, ropes, lanterns and food, but
would leave one item behind – his wristwatch. Instead of
recording glacier measurements, he would systematically
record his perceptions of time passing. He wanted to explore
the natural rhythms of the body untainted by outside cues.
The longest attempts to do this had at that point lasted
only seven days, with both American and Soviet astronauts
taking part in Cold War isolation studies to assess how
people might survive in fall-out shelters after a nuclear
attack. Michel had done the same, volunteering to spend a
week in silent darkness in an experiment at an air force
base in Ohio. Now he wanted to try it out for far longer,
in more extreme conditions.

The authorities were not keen on letting a 23-year-old
embark on such a risky expedition. But Michel was determined and he had form when it came to persuasion, having

convinced a professor at the Academy of Sciences in France to take him on as a geology student when he was only 15. The difference this time was that he was putting his life at risk.

The location Michel had selected for his experiment was the Scarasson Cavern, a cave created from hundreds of horizontal layers of ice, which, unusually for a subterranean glacier, wasn't linked to a glacier on the surface. To reach the ice cave Michel would need to descend a 130-foot shaft, part of which was S-shaped, meaning that if he were to slip on the ice and break his arm it would be impossible to haul him out; a minor fracture would result in death. Assuming he made it safely down into the cave, he planned to spend *two months* down there in complete isolation. He offered to sign disclaimers, freeing the authorities from any legal responsibility for his safety, but they insisted they would still be morally responsible. He was too young, they told him, too inexperienced, and above all too optimistic. Even after he made a year of detailed preparations, some still maintained the whole idea was nothing but a stunt. The turning point came when he delivered a lecture on his previous expeditions to his friends at the Club Martel potholing group. They saw that he was serious and agreed to act as a support team. Still he would need funding and written permission. He made numerous visits to the offices of officials where he would sit and wait, only to be told after an hour or more that the person in charge was too busy to see him. Michel started to feel these office visits might require more perseverance than the expedition itself.

While he negotiated all the bureaucratic obstacles, Michel

theorised about the experiment he was hoping to carry out on his own mind. He speculated that time existed on three levels: biological time, which stretched across many years; perceived time, created by the brain and conditioned by light and dark; and the objective time as shown on a clock. His interest was in comparing the final two. Specifically he wanted to discover through extreme self-experimentation whether humans have an inner clock that somehow synchs with 'clock time' even without any external cues. He also wanted to know how time would *feel*. On past trips underground he had found that time warped. The subterranean world was so absorbing that whenever he returned to the surface he was astonished to discover how much time had passed.

Eventually Michel raised the necessary funds and persuaded the authorities to let him go ahead. Although he would ultimately be all alone in the cavern, during the preparations he had a team of people helping him. For several weeks beforehand his friends from the potholing club stayed with him in his parents' house, preparing equipment and supplies during the day and sleeping in the hallways at night. Meanwhile Michel had been instructed to rest. The team loaded the equipment onto trucks and drove as near to the cave as they could. When the trucks got snowed in they even built a primitive telefiric railway with a cable and brakes for moving the heaviest items. They marched through the snow for hours at a time carrying the rest of the supplies to the mouth of the cavern. They negotiated the difficult descent and ensured that Michel would have all the equipment and supplies he needed. Once the underground camp was set up, two of the men spent three nights in the tent as a trial.

Michel said goodbye to his mother. Again she told him how afraid she was. Again he told her how hopeful he felt. He spent the last night before his descent in a tent at basecamp with the idea that he would feel rested before his descent. Instead his fear kept him awake, and when he emerged from his sleeping bag next morning he had slept so little that his bones ached. As he began the climb up to the mouth of the cavern he was struck by an attack of amoebic dysentery. Things weren't looking good. Still he gave the team strict instructions not to join him underground until two months had passed, signing a statement immediately before his descent, which declared that during the first month no one must attempt to rescue him, whatever the circumstances. Finally he handed over his wristwatch and began his descent with the team. They checked that the tent and camp bed were as he wanted them, taught him how to change the batteries powering the light bulb and the telephone, took some samples of ice from the glacier and then they left. Calls of '*Au revoir*' echoed down the chamber. Michel heard them pull the ladders up. Until they came to fetch him in two months he really was alone. Did his mind or body contain a clock that could judge the passage of time, and would he continue to be able to guess when a minute had passed?

Before we return to Michel at the end of his two lonely months, I want to look at the way we assess durations of a few seconds or more, an amount of time that would seem like nothing to Michel, but which surprisingly counts as

long in time-perception research. Warren Meck is a neuroscientist at Duke University in the USA who studies people
with a skewed sense of time. By examining the cognitive
processes involved in processing durations ranging from
seconds to hours, he has localised the perception of time-
frames of more than a few seconds to an area at the centre
of the brain called the basal ganglia. Until 2001 no one had
any idea that these areas housing a mass of neurons could
be involved in the perception of time. The basal ganglia,
one in each hemisphere of the brain, are deep inside the
middle of your head and they loop around in a curve a bit
like the older kinds of hearing aid which curl around over
the top of the ear. The basal ganglia help control movement
using the neurotransmitter dopamine to put brakes on your
muscles. If you want to sit down you need to stop the
activity in all your muscles, apart from those needed to
maintain your sitting position. Then when you want to get
up the basal ganglia release the brake on the muscles and
that helps you to move off smoothly. Meanwhile a brake is
put on those postural movements that were keeping you
still. Without enough dopamine to power this brake you
would experience both the tremors and jerky movements
associated with Parkinson's disease. You would find it difficult to initiate movement, rather like trying to drive with
the handbrake on. But the basal ganglia are also implicated
in timing events which last longer than two seconds. This
is also something people with Parkinson's disease find difficult. This condition destroys the cells producing dopamine
and the greater the number of cells a person has lost, the
harder they find it to estimate time.

The whole dopamine system appears to be crucial in the perception of time. If you give someone the drug halop-eridol, often prescribed for schizophrenia, it blocks the receptors for dopamine and causes people to *under*estimate the amount of time that's passed, while the recreational drugs methamphetamines (or speed) do the opposite; they increase the levels of dopamine circulating in the brain, which causes the brain's clock to speed up with the result that people then *over*estimate the amount of time that has passed. This might seem somewhat counter-intuitive, but it echoes the process that has been hypothesised to occur when people are in fear for their lives.

EMOTIONAL MOMENTS

The basal ganglia, along with the cerebellum and the frontal lobe, bring to three the areas of the brain we've looked at so far. When you consider their other functions, it makes sense that these areas of the brain also relate to time. But the involvement of a fourth area is more mysterious. A psychologist called Bud Craig noticed that whenever people carried out time estimation tasks in a brain scanner, another area kept showing up yet appeared to go unremarked – an area that processes sensations from the rest of the body. He realised that parts of the body *outside* the brain might have a part to play in time perception.[23]

If it's very quiet and you lie very still in bed at night, you can sometimes detect your heartbeat, without putting a hand up to your chest. Ten per cent of people can feel their heartbeat at any time, particularly if they're lean, young men – it helps to have as little flesh as possible getting in the way. This ability to detect changes in our physiology is known as interoceptive awareness. When I was making a programme about it, I canvassed lots of people to see whether they could do it. None could. As I climbed the many stairs up to my flat I passed the door of the lean young man called Hadley who lives in the flat below. He's used to my asking strange questions for programmes and when I knocked on his door and asked whether he could hear his own heart beating, he instantly tapped out its rhythm on the table. Now no one is suggesting that we keep time using our heartbeats, but interoceptive awareness could play a part.

The area of the brain that interests Bud Craig is called the anterior insular cortex. It allows us to detect how our body feels and is responsible for gut feelings like disgust, or butterflies in the stomach when you're in love – those feelings that are mental, yet on the cusp of physical. This would fit in with research on mindfulness, which has found that during mindful meditation people show stronger activity in the insular cortex. We know that people who are deprived of their senses say that time goes slowly. Could the rate of signals coming in through the different senses, including interoceptive awareness, contribute to the creation of our sense of time?

Craig's model of interoceptive salience can also make sense of Hoagland's experiments with his wife's raging temperature and the effect it had on her perception of time. The awareness that we are detecting warmth, itchiness, aches, thirst, hunger or pain comes through the anterior insular cortex. Craig is proposing that this part of the brain gives us a reading of our emotional state from moment to moment across time – like a string of cut-out paper figures, each representing what he calls 'an emotional moment'. It's as though there is a line of 'you's going from the past, through the present and ahead into the future. The same system that indexes this succession of emotional moments could be used to count time. It could even explain the powerful emotional impact that music can have on a person. The same system that deals with emotions could count rhythm. The beauty of Craig's idea is that it could also explain why fear slows time down. The succession of emotional moments needs to go faster in order to register the intensity of emotions in that terrifying moment. So the clock counts faster, making time feel slower.

It is becoming clear that there are more connections than were once realised between these different areas of the brain that deal with time. All four of the areas I've discussed seem to contribute, which might explain why deficits in time perception after a brain injury are rarely as bad as you might expect; damage to a single, tiny area can transform someone's personality, ruin their memory or cause them to lose ability to speak or even comprehend other people's speech, yet the problems with timing tend to be small and often apply only to one time-frame. This might be precisely because there is no single clock.

So neuroscientists know where the brain counts time, but the mystery remains as to *how*. Does information come in from the basal ganglia and the cerebellum and then somehow the frontal lobe counts up that information, giving a sense of duration, or do we count up these emotional moments as Craig proposes? Either might be true. But there is one problem; with all the developments in neuroscience there is still no sign of this elusive clock to do the counting. There are several different theories about how the brain counts time and I'm going to look at some of the most influential. The chief area of debate concerns whether we measure time using memory, attention, a straightforward clock, a series of clocks or the everyday activity of the brain itself. Any theory needs to be able to explain why it is so easy to trick our sense of timing using the simple method below.

THE ODDBALL EFFECT

Imagine I were to play you a series of seven notes, all of which are identical except for the middle note – so three Cs, then a G, followed by three more Cs. Each note is of exactly the same length, but you would be convinced that the G lasted for longer. Likewise if I flashed up a sequence of pictures on a screen – giraffe, giraffe, giraffe, mango, giraffe, giraffe, giraffe – you would swear that the picture of a mango was on the screen for longer. It is as though time very briefly slowed down when you looked at the mango. This is known as the oddball effect. It is a basic error of timing which we make repeatedly and, simple as the test is, it can give us an insight into how a clock in the mind might work.

One solution that can account for the oddball effect is the minimalist clock model. The idea is that somewhere inside the brain we have a pacemaker that ticks like a metronome, endlessly beating time. This is coupled with a counter that switches on at the start of a given time period and off at the end, counting up the number of beats which occurred. There are various theories about the precise mechanism, the most influential of which is called Scalar Expectancy Theory. The element of expectation comes in like this: if you are played two musical notes and asked to judge which was longer, your internal clock ticks away the milliseconds of the first note and counts them up. Then the second note is played to you. If it is the same length, you now have an idea of how long you would expect it to last. By comparing its actual duration with your expectation, you are able to judge whether it was longer or shorter. This theory can account for the oddball effect. The surprise of seeing a mango instead of yet another giraffe picture wakes you up emotionally which temporarily speeds up the ticking of the clock, leading the counter to count up more beats and giving you the impression that the mango was present for longer. The same happens with the first item in a sequence; its novelty emotionally arouses us very slightly, the clock ticks faster, more beats are counted and so it feels as though it lasted longer.

The problem with this theory is highlighted by studies where people are trained to discriminate between different time intervals. Musicians have the edge on the rest of us when it comes to estimating the passage of time. In Istanbul the Turkish psychologist Emre Sevinc conducted a cunning study where he measured the accuracy of time perception in musicians using pairs of musical notes. People were asked to listen to pairs of notes and to say which pair had the smallest time interval between them.[24] We would expect a professional musician to find such a task easy, and they did, but what Sevinc wanted to know was whether such skills could generalise to other senses. So in the second part of the study the musicians were twice tapped on the hand. Again their task was to identify which pair of taps had the shortest interval of time between them. He was able to establish that skills in time-estimation can generalise to other senses. But, with the shortest time-scales of just 100 milliseconds, the non-musicians were equally skilled (or unskilled – it's quite difficult), suggesting we might need different clocks for different time intervals. Similar studies with non-musicians have found that if you train people in this skill they improve quite quickly, and they can generalise their newfound skills in timing to other senses. But what they can't do is to generalise to other time intervals. Even after the training if you play notes with longer time intervals between them, they are no better at making judgements than anyone else.

So if a single clock can't do everything, does this mean that we have multiple clocks, each dealing with a different time span? If this is the case then somehow our brain manages to synthesise these different processes to give us a sense of time as one seamless stretch. Just as our brain takes the visual input from two eyes and adjusts it to give us a view that appears as one picture rather than two overlapping circles, so our brains make sense of the time cues we're given from multiple processes. Some speculate that we might have the equivalent of a bank of hourglasses in the brain, each pre-set to time a different interval. When you hear a starting tone in an experiment like the one above it sets off certain neural processes that do the equivalent of starting the sand falling through a set of hourglasses. At the next signal the sand stops, and depending on which hourglass has finished, you are given a duration. But then would every possible time interval require its own hour glass? There's no sign of where this imaginary set of hour-glasses might be, yet we can definitely give fairly good timings without a stopwatch and with practise we do get better. When you learn to drive you get used to how long a red light takes to change. Drive in a different country and a differing duration can take you by surprise. I know how long 40 seconds feels from making radio programmes where 40-second clips are so common, and therapists tell me the time-frame that they can best detect is 50 minutes, the standard length of a consultation. Even though they admit to finding therapy sessions with some clients more absorbing than others, they become adept at gauging when the time is almost up.

It is possible that the brain contains no specialised clocks, but instead possesses the ability to gauge size, whether in terms of length of time, number of sounds, distance, area or even volume. Even without rulers or measuring jugs, we are surprisingly good at judging magnitude. If people sustain damage to the area at the top of the head at the back, where the skull starts to curve downwards, they not only have difficulty judging distance, but the position and speed of objects too. This region, known as the parietal cortex, is where we initiate an action which results in a movement. When a baby experiments to see what they can reach, push, lift, fit in their mouth or climb over, they are developing their parietal cortex.

Weber's law states that errors of judgement rise in proportion to the magnitude of the property you're judging. So if you are guessing a distance that's a matter of several metres, any mistake you make will be smaller than if you guess a distance in miles. The Danish psychologist Steen Larsen proposes that since the same seems to happen with time, when we hold a particular duration in mind we somehow have an idea of a distance in time. As with geographical distance, small differences are less noticeable, the longer the time period we are considering. Weber's law applies across species and regardless of the type of magnitude we're estimating, so whether it's babies being tested on their ability to compare the area of two pieces of coloured cardboard or pigeons being tested on the timing of beak-taps to get seed, the same thing happens. This suggests that judgements of size could hold the key to time perception.

So far the competing ideas I've covered for how we count time are these: the presence of a clock or series of clocks, a system based on emotional moments, or something simpler like an ability to gauge size. To work out which is the mostly likely explanation, the number three might help.

THE MAGIC OF THREE

The number three comes up a lot in research on time perception. Spoken language uses a three-second rhythmic structure, with poets often writing three-second lines.[25] It is a time-frame that seems to appeal to us. You see it everywhere from the three-second stings used to break up radio programmes to those irritating start-up sounds on computers. The ethnologist Margret Schleidt worked with four different communities – Europeans, Kalahari Bushmen, Trobriand Islanders and Yanomami Indians – filming their daily lives and then timing everything from the movements of their heads to their feet.[26] She found that in all four cultures handshakes lasted – you've guessed it – three seconds. There seems to be an unspoken agreement on the appropriate time-frame for a handshake. If it is too long or too short you instantly become slightly wary.

It also comes up a lot in experiments on how long a 'moment' might last. In the eleventh book of his confessions, St Augustine declared that the past and the future are mental constructions that we can only see through 'a window of presence'. A long line of researchers have tried to work out how long this 'now' or 'moment' might last. In 1864 the Russian biologist Karl Ernst von Baer proposed that different animals experience different lengths of moments. He defined these moments as the longest possible duration that could still be considered to be a single time-point. An hour is clearly too long, so is a minute, but many in the field feel a moment feels as though it lasts longer than a second, with the exception of the physicist Ernst Mach who wrote in 1865 that the maximum moment was a fraction of that at just 40 milliseconds.[27]

More recent experiments suggest that a moment lasts between two or three seconds, which aligns not only with

what we see in poetry, but also in music, speech and movement. We seem to segment activities into a space of two or
three seconds. Children with autism sometimes find time
perception difficult, and if you play them a musical note
and ask them to play it back to you for exactly the same
length of time, whether it lasted one second or five, they
will almost always play you a note of three seconds.

It is known from many classic studies on working memory
that three seconds is the length of time we can keep something in mind without writing it down or devising a way
of committing it to long-term memory. So if someone tells
you a phone number you are able to dial it immediately as
though you are reading it from your mind, but if you are
distracted (and just pressing the correct buttons on a mobile
to stop one call and start the next would be enough) *or* if
you wait more than three seconds it is very difficult to do.
It is as though every few seconds the brain asks what's new.

One of the most important questions we've been
discussing when it comes to how the brain measures time
is how a mental clock or series of clocks can deal with
different time-frames. Could the same fast pulse in the
brain beat out five minutes as well as 100 milliseconds or
would you need an entirely separate clock? If there are
different clocks for different durations, where are the boundaries? This is where the three seconds come in. Experiments
have demonstrated that the most definitive boundary in the
way we judge different time-frames is between 3.2 and 4.6
seconds.[28]

We saw at the beginning of this chapter that a surprising
number of different parts of the brain are involved in time

perception. Maybe this is because we need to deal with so many different time-frames. We shouldn't expect the sensation of two clicks on a Savart wheel to be measured in the same way as Michel's freezing nights in his black ice cave. The German psychologist Ernst Pöppel suggests two different mechanisms – one for short durations and one for long. Others suggest that there might be a whole series of clocks for different durations, clocks which could sometimes overlap. I like to imagine it like a newsroom where each clock is set to a different time zone, but with a clock for every duration. Yet if this were the case then why can the same time-period feel longer if we are listening to a sound than looking at a picture? Would we need a whole new set of clocks for each of our senses?

It is possible that different areas of the brain measure different time-frames using distinct mechanisms. We know from research on the emotions that the brain is not designed like a neat, ceramic phrenological head divided into different sections for different emotions, but that each emotion uses a different combination of brain systems. Could the same be true of timing? Could the brain use different combinations of areas to assess durations of different lengths?

Perhaps the whole notion of a single clock or a series of clocks is too complex. An alternative explanation focuses on concentration. Just as time flies when you're absorbed in a book, the more complex a task you're given in a lab, the shorter you estimate that time period to be. So if you are presented with a list of words and asked to spot words beginning with E as well as any words that represent animals, this requires two different skills and more concentration than simply looking for animal words alone. So the more that's going on, the faster time seems to go by. The Attention Gate Model is an example of this kind of idea.[29] The theory is that we have a pacemaker which emits an endless series of pulses in the brain, and a gate that allows our brain to count up every pulse that passes through, just like a shepherd counting sheep as they're herded through a farm gate. If you're feeling anxious the pulses speed up, and so more pulses pass through the gate within a given period causing you to believe that more time has passed than really did. In other words time felt as though it slowed down. If you are paying attention to time itself, if you're in a queue for example or taking part in an experiment where you have been told in advance to estimate a period of time, this also causes more pulses to go through the gate, with the result that time apparently seems to pass more slowly. This theory would also help to explain why time passes slowly during periods of depression. While a person is introspecting (or meditating), their attention is turned inward, every pulse of time is noted and the hours appear to go more slowly.

This makes sense, but why should time go faster when you are busy? Perhaps the brain shares its resources between concentrating on the event at hand and timing it, so when you are distracted time flies by uncounted. This is the basis of what's known as the resource allocation or time-sharing hypothesis. And with this theory it doesn't matter what the clock is like – it could consist of a pacemaker or a line of hourglasses or measure the rate at which neurons fire in the brain – the crucial point is that its timing mechanism is disrupted if your attention is displaced. The moment you give people a second task to do the minutes go faster; a watched pot never seems to boil, but go and check your emails and it will be boiling over before you know it. With the Attention Gate Model the more absorbed you are in a task, the less attention you pay to time itself, the pulses slow down, fewer pass through the gate and you believe less time has passed than really has.[30] The clever thing about this model is that it is flexible enough to allow for the influence of emotions, and by now it should be becoming clear that time perception and emotion are strongly linked.

HEADING FOR A PRECIPICE BLINDFOLDED

Jonas Langer, a psychologist at Clark University in Illinois, had an idea. He would build a platform on wheels, stand people on it, blindfold them and then ask them to manoeuvre themselves slowly towards the edge of the stairwell where there was a drop of several storeys. He wanted to know whether they would think time had gone faster when they wheeled towards or away from the edge. As it was the 1960s, when the standards of university ethics were lower, no one tried to stop him. If you look at the illustration below, the platform had handrails to hold onto at the side, but no safety barrier at the front. The volunteers could start and stop the platform using a button connected to a motor that propelled it forwards at a steady two miles per hour while Langer and his colleagues steered it from behind. They were required to drive themselves towards the edge of stairwell while blindfolded from two different starting points – the first 'less dangerous' point was 20 feet back from the precipice, and the second 'very dangerous' point was 15 feet from the precipice. The instructions were to press the button for what they, without counting, estimated to be five seconds. Considering the platform moved at two miles an hour, driving it forwards for the whole five seconds from a starting point of fifteen feet would bring you to within less than eight inches of the precipice. Remarkably eight men and eight women agreed to take part, even after seeing both the drop down into the stairwell and the blindfold. And there was to be no cheating by standing well back on the platform. Each volunteer was made to stand with the toes of their shoes level with the front edge of the platform.

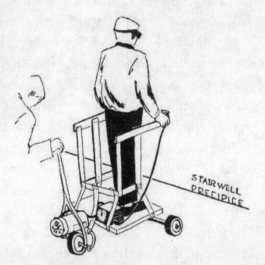

Not surprisingly Langer found that when people were facing danger they pressed the button for a shorter time. He interpreted this as indicating that time slowed down for them due to their fear, so after just 3.6 seconds they thought that five seconds had passed.[31] We know from the last chapter and from personal experience that fear does cause time to appear to decelerate, but with this experiment there is of course an alternative explanation. If you know you are moving towards a precipice blindfolded then it is probably sensible to err on the side of caution and to stop the platform a little early. If they had continued moving for six seconds instead of five, unless the experimenter was fast enough to reach the stop button, they would have driven off the edge and fallen down the stairwell.

Having said that, we know from the many laboratory studies conducted since, that emotions do alter time perception. Just as fear makes time go slowly, so do looking at pictures of mutilated bodies or listening to the sound of a woman sobbing.[32] It seems that when faced with distressing images, your body and mind ready themselves for fight or flight, so the clock goes faster, more pulses accumulate and it feels as though time went slowly.

As we've seen, the passing of time is judged in two ways – prospectively, as it happens, and afterwards, retrospectively. When you judge time prospectively it is easy to see that, as I've been discussing, attention and emotion both play a part; but when you look retrospectively and try to guess how long an event took, it is a third factor that shapes your answer – memory. This difference between prospective and retrospective time estimation is very significant, and one that provides solutions to many of time's mysteries. It gives rise to the phenomenon I've dubbed the 'Holiday Paradox'. This is the common experience of a holiday appearing to go fast at the time, but when you look back afterwards it feels as though you were away for ages. I'll be returning to this in more detail in Chapter Four.

It is clear that memory is involved in time perception, but there are still disagreements over whether we have a separate working memory just for processing time. Do we have a working memory buffer that allows us to hold temporal information briefly in mind in the same way we can retain a phone number for just long enough to dial it? It is possible that a pacemaker counts the milliseconds while complex memory processes allow us to deal with longer time intervals. Studies of patients with amnesia suggest that the processing of time also shares some neurological pathways in common with the creation or recall of certain types of memory. This link between memory and time perception is backed up by the fact that the tranquiliser Valium impairs both memory and time estimation skills.

To sum up so far it seems that we have some kind of clock in the brain which counts time and is influenced by these three key factors: attention, emotion and memory. It could consist of a single clock, or a series of special pulses. There's just one problem: no one can find it.

IS THE BRAIN TIMING ITSELF?

So is it possible that there are no clocks, nor any special pulses for timing? Perhaps the brain is exploiting the activity already going on within it by taking readings of time from the neural networks that are constantly making computations to assess everything from colour to pitch. According to this theory no part of the brain exists in order to actively and deliberately count time. There are no specialised mechanisms for time estimation. Instead time is inferred using the general features of brain circuits that are doing something else – processing spatial information, for example, or recognising a face. A number of neuroscientists are moving in the direction of this sort of idea, and what they want to discover is how the brain might do this. Neurons can produce a steady series of pulses that could be used for timing, but the brain appears to have no mechanism for counting them.

An alternative theory is that we use brain oscillations to time short events. Brain oscillations – the alpha waves of brain activity you can see on an EEG trace – are very short and could lend themselves to the role of the clock. This idea is backed up by the curious experience of going under general anaesthetic. We know that neurons stop oscillating when we've been put to sleep, and anyone who has had surgery will tell you that when they woke up it didn't feel as though any time had passed. This is very different from normal sleep. If the mind does assess the passage of time using these oscillations this could explain why. There's just one problem with this theory. These oscillations last 30 milliseconds, which would imply that the brain counts in 30 millisecond packets. Yet we are able to count durations which are not divisible by 30 milliseconds.

The French neuroscientist Virginie van Wassenhove believes that any set of neurons in the brain has the potential to help us work out timings; it's just a question of turning our attention to it. So the activity is going on all

the time, but it's not until we ask the brain to time the
difference in duration between two musical notes, for
example, that we in effect read off the calculations that we
need. It is a bit like estimating the number of people in a
room – usually we ignore this information coming into our
minds, but if someone asked us to do it, we could. So time
is directly, if not always accurately, 'transparent to consciousness'.[33]

   In his lab in Los Angeles, the neurobiologist Dean
Buonomano is using electrophysiological, computational,
and psychophysical techniques to attempt to discover how
the brain tells the time. On his website you can test your
own skills at temporal processing at tiny durations.[34] The
site plays two pairs of sounds just milliseconds apart and
you have to work out which pair had the smaller gap between
them, a task not dissimilar to that undertaken by the musicians in Istanbul. Buonomano has an explanation for the
finding I mentioned before: that people get better with
practise, until they move on to a different duration where
they find themselves back at square one. Their skills can
generalise to other senses, but not to other time-frames. He
argues that in order to succeed at this task, the brain treats
the sounds like ripples created by a stone thrown into a
pond. The ripples remain for a few moments after the stone
has sunk, like a memory trace of what was happening just
before. When a second stone is thrown in, the pattern of
its ripples is influenced by the waves caused by the first,
allowing the water to momentarily hold a record of both
events. So in the brain, the first tone activates some neurons,
leaving them in a new state, then along comes the second

tone and, because the neurons are in a new state, the response is different. It's as though the ripples left by the first tone provide a new context for the next tone. In the sound task, the brain is able to compare the patterns of activity caused by the first pair of sounds with the pattern from the second, before assessing which was shortest. So we don't need a ticking clock to do the measurement because it is the patterns of activity in the brain themselves that do the timing. He calls this a state-dependent network. The tasks sound easy, but on my first attempt at his test I scored 23/30, not especially good considering that a score of 15 could be achieved through random guesswork alone. Luckily life doesn't require us to perform this exact task, although millisecond timing is so crucial to the production and understanding of language that skills at judging this time-frame could contribute to linguistic ability. Next researchers are hoping to discover whether deficiencies in timing might underlie conditions such as dyslexia. This could explain the unusual relationship with time experienced by people like Eleanor, who is persistently late because she has no accurate sense of time passing. Is it possible that it is the precise timing of the movements of a pen on the page or of reading a series of letters that allows us to write and read accurately?[35]

Experiments using three sound tones lend weight to the idea that we don't need a specialised central mechanism to estimate time, that instead we read it from the activity of the neurons that are there doing other jobs. Volunteers are asked to judge the interval between two tones while ignoring a third irrelevant tone that is played first. If the brain has its own stopwatch this shouldn't be a problem; you just reset the stopwatch after tone one and time the interval between the second and third tones. But this isn't what happens. The third tone does confuse people, suggesting that neural activity not specifically designed to measure time does the timing and that this timing is thrown off course by the introduction of an extra tone. This makes the system imperfect, but the advantage is its flexibility. In theory the system can time anything that's happening, whichever sense it comes from. Crucially a third note is not distracting if the pitch is different. This makes me wonder whether different sets of neurons are used to do the timing for every different note played.

The same David Eagleman who throws people backwards off buildings has another idea which, like Buonomano's theory, relies on the idea that our brain cells possess inherent properties of timing. When you look at a picture it takes a certain amount of energy for the neurons in your brain to identify what you are seeing. Think back to the giraffe/mango task, where people who are shown a series of pictures of giraffes with a surprise mango in the middle insist that the mango was present for longer than each giraffe. Registering the giraffe the first time uses a certain amount of energy. On seeing an identical giraffe picture the brain doesn't need to waste as much energy considering it. Eagleman's theory is that our sense of duration is based on the amount of neural energy used. So the first picture of the giraffe would take more energy and therefore seem longer, while subsequent giraffe pictures use less energy and seem shorter. Then up pops the mango. It's new and requires more energy for the brain to register what it is, so it appears to be present for longer. In terms of the evidence for this idea, it is true that rates of neuronal firing do increase when there's a new picture and drop off when the pictures repeat. Whether this is exactly how we calculate timings remains to be seen, but it is certainly a plausible idea. We know that novelty does play a part in timing, even at longer durations. If you arrive in a new city and walk from your hotel to a restaurant, the processing of all the new sights and sounds will use up a lot of neuronal energy, which will give the impression that the journey has taken a long time. The walk back, along a route now more familiar, will seem shorter.

The idea that neuronal activity itself might be used to measure time could also explain the difficulties that people with schizophrenia can have with time perception. Unlike voice-hearing or delusional thinking, this symptom tends to be less well-known, but some people find they are no longer able to view the present while simultaneously remembering the recent past and anticipating the future. The philosopher Edmund Husserl believed that holding these three time-frames in mind was essential for consciousness and for giving us a firm sense of reality. In schizophrenia this can be disrupted, making time feel unreal. People with schizophrenia find it hard to spot the oddball in an experiment or even to detect flickering lights. Their neural responses suggest that everything they see seems fresh and new. Usually when you show someone the same giraffe again and again their neural response diminishes, but not with people with schizophrenia.[36]

We can predict the timing of everything from the swinging of a pendulum to a car door closing without trapping our fingers. We don't even notice these small timing judgements that we make hundreds of times a day. But imagine how unsettling it would be if they started to appear to be out of kilter. Add to this the disturbance of your thoughts. If you have lost all cues to temporal reality and have no way of ordering your thoughts chronologically to identify which are memories, which are daydreams and which is the reality here and now, it is no surprise that a psychotic episode can feel terrifyingly disorientating. Philosopher and neuroscientist Dan Lloyd goes as far as to suggest that a disorder of temporal perception might even account for some of the symptoms clustered under the diagnosis of schizophrenia. This makes sense. I've already mentioned the influence of dopamine in time perception and one of the theories for the cause of schizophrenia, 'the dopamine hypothesis', implicates this neurotransmitter. It is possible that dopamine in effect sets the mind's clock, dictating the rate of the pulses, and that some symptoms of schizophrenia could result from a disruption to the clock.

Eagleman's theory can also explain the stopped clock illusion. The initial tick seems longer because it's the first time your brain has registered the movement of the second hand, then the neuronal firing and the energy used drops off and so does the perception of time passing as the second hand continues on around the clock face. Likewise a bright light that's turned on momentarily appears to last longer than one that is more faint and an interval filled with a complex piece of music seems longer than one with a simpler piece. Is this because we are timing them via the energy used to process them?

I realise I have covered a lot of different theories here. My best guess, based on the available evidence, is this: that pulses which are already being used for other purposes are measuring time in our brains. They might be ripples, they might be energy packets, but, whichever they are, when we turn our attention to time itself, they speed up. This acceleration, which you might recall from the discussion of the sheep going through the gate, gives the impression that time is dilating. Extreme anxiety also speeds up the pulses, so while Chuck Berry desperately tries to save his own life the pulses get faster and time gets slower. To estimate a length of time, we use the dopamine system along with combinations of those four crucial areas of the brain – the cerebellum, the basal ganglia, the frontal lobe and the anterior insular cortex – depending on the time-frames involved.

Once again this illustrates the central theme of this book – that we are creating our own perception of time, based on the neuronal activity in our brains with input from the physiological symptoms of our bodies. This could seem like a reductionist explanation, that time is simply chemical, something created by neuronal activity in concert with the dopamine system, but these neuroscientific explanations should not reduce the importance of our subjective experience of time. For Chuck Berry or Alan Johnston or even Michel Siffre lying in his cold sleeping bag in his ice cave, the pulsing of neurons meant nothing. It was the experience that mattered and this is the part we can change. We have

multiple skills when it comes to considering time. We can
mentally throw ourselves forward into the future or back
into the past. We can imagine situations in the future that
we have never witnessed, we can put events into chronological order, detect rhythm in music, speak, catch a ball,
run for a train, cross the road and all without needing any
awareness of what's happening in the brain.

Yet this reality we create for ourselves can easily get
disturbed. Eleanor finds it hard to judge time without a
clock. How much harder would it be with no daylight and
no one else to ask?

OPERATION TIME

For two months, or 1,500 hours, Michel Siffre had lived in
total isolation underground in the French Alps with no idea
whether it was night or day. He allowed his body to tell
him when to rest and went to bed whenever he felt tired,
describing this body-dictated sleep as more perfect than he
could ever have above ground. He ate when he was hungry.
But he soon lost his appetite. The one advantage of the low
temperatures was that his food stayed fresher for longer
than he had expected, but he was no cook and his attempts
to make rice pudding went so wrong that he had to open
a tin of pineapple chunks to take the taste away. In the end
he found he only really enjoyed bread and cheese. Each
day he read, kept a diary and noted down physiological
measurements from the electrodes fixed to his head and
chest. The experiment he had dreamed of conducting for
so long was going well, but he was becoming increasingly

miserable. His mattress was made from a thick piece of sponge, but because it lay on a floor of ice in below-freezing temperatures, the bed was often damp. His feet were permanently wet and the air permanently dank. His clothes never dried out overnight, so he would put them back on next day and shudder at the bitter wetness against his skin. After spending so many hours sitting down each day he developed back pain, but was determined not to take painkillers in case they interfered with his physiological experiments.

Michel found himself passing the hours by contemplating another time entirely – the future. He tried to find ways to entertain himself; his version of quoits involved attempting to throw sugar lumps into a pan of boiling water. A record player had been lowered down the shaft for company, but Beethoven and Mario Lanza weren't a success. 'The symphonies that had once charmed me became merely chaotic noise. And the popular songs sung by the best café singers seemed only to increase my feeling of loneliness.' He was so lonely that the only thing he describes with any real pleasure in his diary is a spider he captured and kept in a box as a pet. He talks of often looking at her and feeding her tiny amounts of food and liquid.

However, despite the wet conditions and his increasing hatred for the yellow lining of his tent, he became so fond of his makeshift home that he began to spend more and more time in bed, leaving the tent as infrequently as possible. When he did venture out into the cave to take measurements he loved to look back and see his cosy, freezing home glowing in the dark. He soon lost interest in keeping the cave tidy and allowed the rubbish to pile up outside the front of his tent. The low temperatures meant that the food was slow to rot, but he did notice mildew growing on an apple core and, ever on the look-out for an opportunity to experiment, he left a line of apple cores so that someone could check on the mildew's progress the following year.

With no daylight Michel developed a squint and found it increasingly hard to distinguish green from blue. He didn't feel claustrophobic, but towards the end of his underground stay he was experiencing dizzy spells and afterwards doctors confirmed that his body had entered a state they called 'incipient hibernation'.

Throughout his stay two members of the team remained above ground at the cave's entrance, sitting in baking sun during the day and lying in freezing temperatures at night. They were forbidden from contacting him lest this give him a clue as to the time of day. Instead, a telephone line had been rigged up which linked him with the surface and he called them whenever he woke, ate or was planning to sleep. They were under instruction to keep a record of the precise hour at which he phoned, but not to reveal it to him. By the second morning Michel was already two hours off-kilter. Within a week he was two days behind reality. Within 10 days he thought night was day and even noted in his diary that with their cheery 'Hellos' the team sounded as though they'd been up for hours. In fact he had woken them yet again in the middle of night.

During each phone call he took his pulse and counted from 1 to 120 at the rate of one digit per second. But here something extraordinary happened. He thought the count took the two minutes that it should, but his colleagues with the stopwatch knew it was taking five minutes. Life without day or night had skewed his mind time. He had lost any accurate sense of the passage of hours or minutes and found he couldn't even guess how long his phone call to the surface had lasted. Initially he used his Mario Lanza records to judge short periods of time, but soon, 'The beginning and the end of a record blend and become integrated in the flood of time . . . Time no longer has any meaning for me. I am detached from it, I live outside time.' Time had become something he could no longer judge, something he found strange.[37] He was undoubtedly bored and lonely, yet found that although each day felt endless, when he looked back he thought it had lasted far fewer hours than it really had. This is a common paradox of time. Yet it was passing even faster than he realised. He eked out his cheese rations to make it last the whole two months, but he was so wrong about time that in fact he needn't have deprived himself.

He did have a suspicion that he might not have the right date, that he might be a few days ahead, but insisted he couldn't possibly be behind. Then the team suddenly announced that the experiment had come to an end and that it was already 14 September. He was astonished. He thought he had 25 days to go. But the discovery that he could now leave the damp cave behind and emerge into the sunlight did not bring him joy. Instead he was confused. He felt he had lost his sense of reality and as a result had lost 25 days. Where had that time gone? He felt cheated of his memories.

Then time warped once again. Although he was expecting to stay for almost another month, as soon as he discovered that the team were on their way down to fetch him, time

felt unbearably slow. Even in the last few minutes before
their arrival he wondered how they could be taking so long.
He had always known that once the team arrived they would
have to spend one more night underground getting everything ready for the ascent, but now he felt too impatient
to wait. And he was afraid. He found himself overcome
with the fear that having survived this long, he might die
right at the last moment. The sound of every tiny rock fall
or crack in the ice made him flinch. Finally his friends
reached him and he felt calmer. They were disgusted by
his rubbish tip, which was by now waist-high, but relieved
to see that he was okay. At the last moment he delayed
leaving. He knew the press was camped out on the surface
waiting for his glorious arrival, but he continued collecting
samples from inside the cave until his colleagues told him
he really must stop.

The journey back to the top was tough. In his weakened
state he had to be winched up in a parachute harness, but
even so he blacked out and almost gave up when he had
to climb using his own strength through what they called
the cat hole. They covered his eyes to shield him from the
daylight. He blacked out again and was rushed to a helicopter, but not before his friend Anne-Marie had held some
fresh violets to his nose for him to smell. This was to become
a very strong memory for him; the first nice smell he had
experienced in two months.

Some claimed the whole operation had been nothing
more than a publicity stunt and that the phone contact
meant he wasn't truly living in isolation, but most accepted
that at the age of 23 Michel had founded the field of

chronobiology – the scientific study of the effect of time
on biological rhythms. His experiments demonstrated for
the first time the existence of a body clock that can function independently of light and dark. Before Michel's
experiment no one knew how the body's rhythms worked,
but analysis of his sleep and wake cycles revealed that
regardless of the time of day, if a series of sleep and active
periods were added together they always came to 24 hours
and 31 minutes. This is the one clock in the body that we
can precisely locate. It is in a part of the hypothalamus
gland at the base of the brain called the suprachiasmatic
nucleus. The neurons here oscillate constantly, providing
a rhythm of just over 24 hours which is corrected by
daylight.[38] Because Michel had no daylight he began doing
what is known as free-running and each day he became
another 31 minutes out of synch. Eventually he became
so off-kilter that he was sleeping during the day rather
than at night, yet his body was keeping to a surprisingly
regular routine.

For his mind it was a very different matter. His perception of time had warped to the extent that every hour felt
three times shorter, despite his loneliness and boredom. He
would stay awake for an entire day and evening and believe
he had only been conscious for a few hours. He was taking
to extremes the disruption in time experience by Mrs
Hoagland in her fever. In one sense time had gone quickly;
he was at the end of the experiment before he knew it. But
in another he had slowed the pace of time in his own mind;
time had expanded for him.

After his 1962 expedition, Michel spent another 40 years

researching time perception, continuing to use caves rather than laboratory isolation chambers for the simple reason that some people are so fanatical about caves that they are prepared to volunteer to spend a month entombed. Sealed laboratories don't seem to inspire the same passion. The French department of defence funded Michel's research in the hope of finding a way for submariners to sleep just once every 48 hours. But after the end of the Cold War he found it harder to secure funding, and now Michel believes it is only mathematicians and physiologists who will be able to take the subject further. Now in his seventies his love of caves continues. Naturally he celebrated the millennium underground and like any good Frenchman took champagne and *foie gras* with him. But having been there for some time beforehand, on the big day his sense of timing went askew and he toasted the new millennium three and a half days late.

MONDAY IS RED

3

'BASICALLY I SEE time as if I'm sitting facing a wall-paper pasting table. I sit near to the right-hand edge, turned slightly sideways so that I look across and also back down the table. The paper starts close to my right hand (the present) and stretches back to the left at the extremity of the table. Ancient time is not actually on the table – like wallpaper, it is in a roll that has tumbled off the far end. I view historical time from an English perspective, in terms of reigns of monarchs. From the far left of the paper to about halfway along the table it is an actual genealogical table, showing Normans, Tudors, Stuarts etc. This stops at about 1800, with a long line running across the paper at an angle of about 15 degrees, to the far side, at 1900. There are two big, rectangular sleepers placed across the track, marking the First and Second World Wars. The far edge of the paper is the English Channel and everything beyond that is "abroad". The map continues across the world, curving away as if on a giant globe.

I notice that Burma has a sleeper marking the expulsion of King Thibaw in about 1885, in the spot otherwise reserved for Queen Victoria's family or the declaration of the German empire.

'Days of the week are like a simple set of five straight dominoes, with two doubles turned sideways at each end for the weekends. Again I jump back when nearing the end of the week. In this map, time moves from right to left. In all others it goes from left to right.'

These are the words of a radio listener called Clifford Pope. I wonder whether they make any sense to you. Or how about this written by another listener – 73-year-old David Williams?

'I see the year laid out in a roughly elliptical shape viewed from above. It's March now, so I'm looking down on early March, with the scene curving away to the left towards April and May. Way over to the far left-hand side of the ellipse I can see August and September. Distant history begins at some point almost out of sight, far away to the right, but approachable by turning away from the year-ellipse just about where April is. That more or less coincides with the earliest nineteenth century.'

Although you'll visualise it in different ways, research suggests that for around 20 per cent of the readers of this book, the idea that you can see time in the mind's eye will make complete sense. And for the other 80 per cent, strange as it might sound, you may 'see' time to a greater extent than you think you do. In the meantime, bear with me.

As I said in the last chapter, we still don't have a comprehensive theory of how people track time. And there is no organ solely for sensing time. But as I'll be demonstrating in this chapter, the ability to picture time in space is particularly important in helping us to create our own perception of time. Not only that, but it can affect the language we use and it paves the way for something that no other creature can experience – mental time-travel.

Through my own research it is clear that some ways of visualising time are more common than others, something backed up by other studies in this field.[39] With the help of listeners to *All in the Mind* on BBC Radio 4, I have analysed the ways in which 86 people visualise time in space. Some sent me long descriptions, including diagrams, and while many commented that they had always assumed that everyone saw time in space, others, like Simon Thomas, thought it was an idiosyncrasy peculiar to them:

'Until I listened to your programme I thought it was just me! I've done it all my life and as a child I assumed that everyone else did it too, until I tried discussing it with a few friends and ended up feeling a little stupid. Since then, and because I find it quite difficult to explain anyway, I've largely kept it to myself.'

Whether or not people feel able to discuss the pictures they have in their head, most seemed almost fond of these mental images. They even commented on how much they had enjoyed the challenge of trying to draw or describe them.

The ability to see time laid out in space is considered by many to be a type of synaesthesia, the condition where different senses appear to blend in the brain. The most common form of this condition involves associating colours with letters, numbers, names or days of the week. In my small survey, I charted the colours people gave for the days – everything from white marbled with orange for Tuesday to a beigey mustard for Friday. There's an intriguing specificity about the precise tones and shades people ascribe to the different days, but I wanted to look for patterns. Is it possible that these colours are based on nothing more than cultural associations? To me Monday is clearly red. Is that because it's the start of the week in Britain and so it's a busy day that stands out? Perhaps most people who see colours for days in predominantly Christian countries see Monday as red. Not the case. It seems my fellow Britons are just as likely to see Monday as pale pink or light blue. To anyone who doesn't see the day in colour this might all sound very strange, and people often assume that we're inventing it, but countless studies have demonstrated that these associations are stable over time and too detailed for people to memorise; test me now and test me in five years and I'll still insist that Mondays are red.

Synaesthesia is now a well-documented phenomenon recognised by the scientific community. Rarer forms can even involve tasting shapes. I've never forgotten reading about a man who insisted that the chicken tasted too pointy, or meeting a woman who described seeing elaborate patterns on hearing certain types of music. When I played her some guitar music she spoke of seeing a quadrant divided into brown, blue, green and navy blue with a river of colour curling down like a plait from the top right-hand corner. As I mentioned above with my red Mondays, the extraordinary thing is that if you play these people the same piece of music or give them the same food six months later they will describe exactly the same associations, and you can test them on multiple stimuli that would be impossible to memorise. Synaesthetes are not making it up. They are experiencing *real* sensations – as evidenced by the areas of the brain that light up in brain-scanning studies. So when the woman mentioned above hears guitar music the areas of her brain relating to colour vision are activated.

No one knows the exact cause of synaesthesia, but one theory points to the richness of connections in the brains of newborn babies. In our first months, the mass of sensations pouring into the brain are not all channelled down specialised pathways. It is as if the brain is a tangled jungle – sight and sound and smell and taste are all mixed up and hard to differentiate. Then at about four months a process of pruning starts, with all the vines and creepers cut back, leaving only separate branches for the senses. Out of confusion, clarity emerges. However, according to this pruning theory, for synaesthetes a few of these jungly connections somehow remain intact, with the result that they continue to experience some crossover of the senses. This idea is supported by the fact that many synaesthetes find that these connections get weaker as they get older. To extend the forest metaphor – while not all the entwining is hacked away, it withers back gradually as time goes on. I've found this happening myself. The colours I associate with people's first names are so much fainter now.

All of this gives credence to the pruning theory, which remains the most compelling explanation we have so far for synaesthesia. But just one thought. The association of letters and colours is the most common form of the condition – yes, sadly, I'm only a common synaesthete – but it appears to present something of a challenge to the strongest theory. For while newborn babies experience a great deal in their first months, they don't see a lot of the alphabet.

MONTHS GO ROUND IN A CIRCLE

The specific form of synaesthesia relevant to time perception is the phenomenon of 'seeing time in space'. As many as one in five of us perceive time in this way. If asked to explain what they mean by 'seeing time in space' people often resort to drawing a diagram. I can understand why, as this is a concept that is hard to describe in words. I will do my best, however, with a few diagrams to help out along the way. For the sake of clarity I will use the somewhat jargony term 'spatial visualisation'. Before I go on, I should say that there has been debate among researchers about whether the spatial visualisation of time, although a genuine phenomenon, really counts as synaesthesia. I believe it does as it exhibits the two key features of the syndrome – the ability to describe perceptions in the same terms both automatically and consistently over many years. It is also the case, as I will show, that the way people visualise time spatially seems to develop during childhood.

In all the radio programmes I've made, I've never known a topic elicit such a big response from listeners as time-in-space synaesthesia. People seemed thrilled to learn that other people also visualise time spatially. Thrilled and liberated. One listener, Sara, told me that the discovery that her experiences were part of a recognised phenomenon was like having a switch turned on in her mind. She had tried hard to suppress her sense that she saw time in space, but now she could let go, 'I suddenly saw the week spread out around me. It was a feeling of relief, of putting things back into place.' Some people told me they were convinced that it was only through seeing time in space that they had any conception of time at all.

Again, I ask for patience for those of you who find all this talk deeply weird, because although it's true that it is only a minority of people who visualise time spatially, the phenomenon can shed light on how mental images of time affect everyone's thinking. Before reading the results of my analysis of the qualitative descriptions below, take a moment to think about how you might visualise time if you were forced to. Obviously time is not a visible concept, but if you had to draw a diagram of time how would you do it? Are the next few weeks laid out ahead of you? If you think about the two world wars do they occupy different places in your mind's eye? Can you look back down the decades? Where is next Tuesday?

Of the 86 descriptions I analysed, the months of the year were the units of time that people were most likely to see laid out in front of them. But this 'laying out' took different forms. Two-thirds of participants who visualised the months in space described a circle, a loop or an oval, with a smaller minority seeing a wave or a spiral. It's not surprising perhaps that some sense of circularity was a strong feature, when you consider the repeating nature of the months through the years of life. We all know the feeling – described by Flanders and Swann in one of their comic songs – of getting to the end of another year and thinking it's 'bloody January again'. The year has completed a full circle and is back where it started. By way of contrast, time/spacers are likely to see the non-repeating decades as consisting of jagged shapes or even zigzags – but more of that later.

Back with the circle of months: July and August were often seen as elongated in comparison with other months, perhaps reflecting the long school holidays experienced in their youth by this British sample. It is also common to report a gap between December and the following January, as though there is a natural break in the cycle at the turn of the year. The strong influence of traditional ways of organising time is apparent here. Only six people described the layout of the months in shapes involving straight lines, with either squares, ladders, rulers or parallel columns of months. One participant commented that she had feared she might lose her mental picture of time on retirement after a working life where time was ordered, but the image was so strong by then that in fact it remained.

The direction of the months in the circle also threw up some curious findings. You might expect the order of the months to mirror that of a clock. But almost four times as many people saw the months of the year going anti-clockwise rather than clockwise as you might expect. One person even mentioned that as a new teacher she had spent the weekend making her pupils a chart of the months of the year. She put January at 11 o'clock, February at 10, and so on, until November was at 1 o'clock and December at 12. The diagram below makes it clearer still.

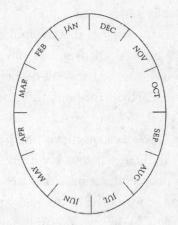

She proudly put it up on the classroom wall, but by Monday lunchtime the head teacher had made her take it down saying the months were the 'wrong' way round. 'Wrong' in what sense? Because January is normally thought as of the first month perhaps, and November the eleventh? Or was it that this circular calendar went in the opposite direction from the clock on the wall? Or maybe the head teacher saw month-time as going clockwise rather than anti-clockwise. As ever, different people will 'see' it in different ways.

It is interesting that people's images of time in space allow for the idea that time is infinite – with no beginning or end. They do not seem to view life quite like the eighteenth-century poet Dryden, who famously described it as 'a crack of light between two eternities of darkness'. True, and unsurprisingly, their own lifetime stands out in bold, but it is not framed in black – or even framed at all. Rather the picture fades away at the edges – like an ink spot on blotting paper perhaps. The time close to before they were born is more distinct than early history where the picture gradually fades away.

This image seems to suggest that the individual perceiver is at the centre of time. But it is not as simple as that. For some people claim to zoom in and out of their time pictures as if on a Google Earth map, homing in on the detail of an individual day and zooming back out to see the span of the centuries. As the decades move on, people see their position on their mental map move with the times.

The pictures people describe really are extraordinary. How about a year represented by an oval with tentacles or by the distinct outline of the state of Zimbabwe? Remember, also, that we are talking about time in *space*. These 'pictures' are not necessarily – indeed almost never – *flat* or viewed head-on, as it were. They are not year planners on an office wall. They are not drawings on a flip chart or even PowerPoint slides. Nothing that simple. They are in 3D; they exist not just 'in front' of a person, but 'around' them. For example, people sometimes describe their picture of time in space as something that circles their bodies, like a sash around a beauty queen – a phenomenon that has also been observed by the psychologist Jamie Ward in his research in this field.

When it comes to how different people visualise the days of the week in space, there is more variety than with the months of the year. A small minority see flattened ovals; others see variations on a horseshoe shape, a semi-circle, or even a curve which Escher-like links back from Sunday round to Monday again. Others see a grid, a piano keyboard or steps; and several described dominoes, lined up one behind the other, a feature that appears with the decades too. One distinct element was that the weekends were some-times delineated in a special way, raised up as steps breaking up a pathway or, like Clifford Pope at the start of this chapter, as dominoes turned sideways.

This is how Roger Rowland sees the days of the week. The weeks stretch away into the future and the weekends are enlarged rectangles.

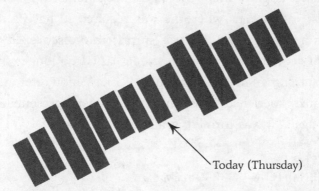

Today (Thursday)

Perhaps surprisingly, very few of the layouts look like diaries or calendars, but some do seem to resemble pictures people might have seen in books or on the walls in primary school classrooms. This is important. These ways of visualising time in space – which seem to be very important to us and useful too – seem to be formed in childhood. I can remember a poetry book which featured a poem about the months of the year, set out on the page in an oval with appropriate illustrations for each month – lambs for May were skipping down on the left, while squirrels buried nuts in October up on the right. Over the years, the illustrations in my mind's eye have probably changed a lot, but this way of laying out the months has remained the same. This one poem, recalled from childhood – and vaguely at that – might have shaped the way I visualise the year in space for the rest of my life.

One participant in my study recalls visualising the day in such a way that the mornings took up far less space than the afternoons. This mental picture did not reflect the layout of any diary or timetable, but rather his experience of short mornings at playschool where he enjoyed himself, followed by long afternoons where he was required to have a nap when he wasn't tired. He was surprised to discover at the age of six that the afternoon in fact lasted no longer than the morning. In this case the spatial visualisation referenced personal experience rather than any external representation. And again, it was formed at a young age.

THE MILLENNIUM PROBLEM

Not so long ago, the millennium played havoc with my spatial visualisation of time. I had my own millennium bug and it seems I wasn't alone.

I transport you back to 1999. I am sitting at my desk contemplating time. For me, the decades of the twentieth century went down a vertical line to 1900, at which point a right angle was turned and the units of time changed from decades to centuries. So after the turning point of 1900, I 'saw' the centuries stacked neatly behind each other – like books on a trestle table – with the decades hidden as if they were chapters unseen behind the book covers.

As I have already mentioned, psychologists who have studied this area are convinced that the ways in which time is visualised in space are constructed in childhood and remain relatively unchanged after that. This might explain why people like me who see time in space 'illustrate' the units of time such as decades and centuries with images gathered when we are very young. So for me, back at my desk in 1999, the decades of the twentieth century were variously illustrated in my mind by actual memories from my lifetime – the 1970s was childhood, the 1980s my teenage years – or by remembered images from TV or cinema – the 1940s, the War, the 1930s, the Depression. With the earlier centuries the differentiating images came from remembered illustrations from books or dramas – the nineteenth century was children up chimneys, the eighteenth Jane Austen dresses, the sixteenth century Henry VIII standing boldly hand on hip.

Now, in a sense, this way of seeing the past is very obvious – even those of you who don't think you see time in space will have images that come to mind when you think of certain periods in history. It is also probably true that you have no such images for, say, the 2070s. But for me, in 1999, my millennium problem wasn't to do with my inability to come up with a visual representation for the new century. Rather my whole way of neatly ordering time in space broke down at 2000. When years such as 2003 or 2009 were mentioned, a mere four years or a decade hence, in my mind they had no 'place' – they were grouped vaguely in a gauzy mush. In short, I couldn't visualise them in space.

And this problem was very precisely centred on the year 2000. It was not just a matter of finding it difficult to visualise the future. In the 1970s I could 'see' the 1990s as a distinct decade in my mind, in a settled place on the twentieth-century line, even though of course I could not 'illustrate' it with images from my lifetime or anybody else's. But the problem was much more profound when confronted with the imminent turn of the millennium.

I was influenced no doubt by the weighty significance the year 2000 was given in the countries that share our calendar. Even when I was a child the millennium was discussed as an important turning point. Then there was the issue of the spatial visualisation of numbers heavily influencing the spatial visualisation of time. In terms of numbers, 1999 turning into 2000 is a big deal. But this was not the whole explanation. My birth date is important here too. Born in the 1970s, I constructed my way of seeing time in space with the subsequent two decades neatly ordered in my mind. Looking back, there was a break at 1900 where my spatial visualisation of time broke into bigger units of whole centuries that were organised in a different way: my books on a trestle table. But looking forwards I didn't go any further than the year 2000. This surely represented a similarly big break, *except* that the Noughties, the 2010s and hopefully a few decades more after that were going to be decades I lived through. It wasn't appropriate therefore to 'see' all the time after 2000 as a single block, but there was nowhere left in my mental picture for the individual years or decades beyond 2000 to go. At some point, time after 2000 would get organised in my mind – as indeed has happened – but until the 2000s actually arrived I could not visualise how it would happen.

Writing this down I have to admit that it sounds very strange, but it wasn't only me who suffered this disturbance in my spatial visualisation of time. The millennium disrupted many well ordered time maps, both among participants in my small study and people whom Jamie Ward has researched in his work at Sussex University.

Clifford Pope was one of many who had nowhere for the year 2000 'to go':

'The point marked by 2000 is curious. For a few years the period after 2000 looked like a short waving piece of string. It had no precise location on the wallpaper. Then in about 2005 it seemed to fix itself, and now very definitely makes another 90-degree turn and runs on to the right along the edge of the table. For some time I continued to view it receding away from me, but more recently I have noticed that I appear to have shifted my viewing position, and seem to be located on the end of this new line, looking straight back along the edge of the table. This is not yet permanent – quite frequently I revert to my old position, and see all the twenty-first century as the future.'

For me, Clifford's description makes sense. And yet of
course it is weird too. 'Short waving piece of string'; 'no
precise location on the wallpaper'; 'looking back along the
edge of the table'. You might well ask what string, wallpaper
or tables could have to do with the perception of time. And
it is not only this disconcerting twenty-first century that
brings out the curious. My vertical lines for the twentieth
century are dull compared with the elaborate descriptions
some people sent me. Centuries are depicted as undulating
ribbons, rows of columns or coils; decades are seen as
towers, bridges, conveyor belts, hawthorn hedges or even
stretched elastic bands. The breaks between centuries and
decades are pictured as dividing doors in corridors, hurdles
on racetracks or sharp zigzags.

This is how Lisa Bingley sees the decades, although she
stresses that you need to imagine it in 3D.

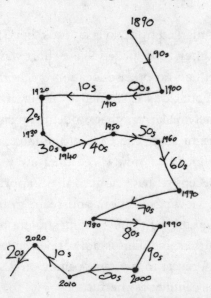

This compartmentalisation into decades is intriguing and appears to be a relatively modern phenomenon. Name a recent decade and the chances are you will be able to summon up an instant image which represents it (and which is commonly shared) – post-war austerity for the 1950s, free love for the '60s and bankers drinking champagne in the '80s. To re-use an obvious metaphor: the book of the twentieth century is written in our minds in neatly divided decade chapters with the two World Wars slightly disrupting the pattern. Yet these neat divides are not perceived as time 'goes along'. Even with that momentous turning point the year 2000, what changed? From one second to midnight on millennium eve to the first second of the first day of the century – nothing at all really. Yet, even now, in the early 2010s, we are probably all beginning to make an ever clearer distinction in our minds between the 1990s and the Noughties, between the decade of growing optimism and prosperity following the collapse of the Berlin Wall in 1989, and a darker decade shaped by the events of 9/11.

We've become very attached to using decades as units for organising time. Yet according to the historian Dominic Sandbrook, in Britain, for example, in earlier centuries decades were barely mentioned as time was divided by the reigns of different monarchs. Of course, both the reign of Richard II and the decade of the 1920s are arbitrary units of time. And the further into the past they recede, the less utility they possess for helping us to organise time in our minds. I'd bet that for you Richard II (1367 to 1400) has been swallowed up into 'the Middle Ages'. In the same way, the 1920s will for people of future centuries be lumped together into a general representation of the twentieth century. Just think: these people in their space suits will be no more likely to differentiate between an Edwardian gentleman and a Teddy Boy than we do between styles of armour in the 1310s and the 1350s. And they may well be confronting their own millennium problem – where to put the year 3000!

COLOURING-IN HISTORY

You'll recall that, for me, Monday is red. For some people whole decades or centuries have their own colours and it needn't be a solid block of a single colour. One person wrote to tell me that the even years are in the light and the odd numbers in shadow. For many, gaps in historical knowledge are often just left dark, and when there is colour or light it doesn't necessarily correspond with the prevailing mood of that time in history. For one listener the First World War took place in a decade bathed in sunshine. For others the 1940s are purple and the Elizabethan period is midnight blue. I particularly like this description from Katherine Herepath, who tells me she loves history:

'I see the last two millennia as laid out in columns, like a reverse ledger sheet. It's as if I am standing at the top of the twenty-first century looking downwards to 2000. Future centuries float as a gauzy sheet stretching over to the left. I also see people, architecture and events laid out chronologically in the columns. When I think of the year 1805, I see Trafalgar, women in the clothes of that era, famous people who lived then, the buildings, etc. The sixth to tenth centuries are very green, the Middle Ages are dark with vibrant splashes of red and blue and the sixteenth and seventeenth centuries are brown with rich, lush colours in the furniture and clothing.'

These are not just pretty pictures; they are useful too. They allow people to hold knowledge in mind and put some kind of order onto the thousands of historical facts they've picked up. The memory champion Ed Cooke employs visual imagery of time deliberately. He suggests that to remember everything in your diary you imagine each day as a different object. If Monday is a car, then each hour of the day is represented by a different part of the car. Then you mentally place your appointments in the appropriate section of the vehicle. So you might picture your dentist perched on top of the steering wheel to symbolise your 10 a.m. appointment, while a tiny version of your boss is trapped inside the headlight for your 2 p.m. meeting. He is employing visualisation as a deliberate strategy, but it involves creating the basic scaffolding for the appointments to hang off. Those with time/space synaesthesia have a distinct advantage. The scaffold is already there. There is a ready-made picture in mind for them to use should they choose to enhance their memory for events in time. One participant in my study told me her pictorial view of the past had been essential for memorising dates for her history and law degrees. But the ability to visualise time in space is not just useful for making sense of history.

Some people sent me accounts of using spatial visualisation of time to plan ahead. Others used the vividness of the mental pictures to help with memory feats that have no association with time; one had such a clear view of the months of the year forming a circle that in her mind's eye she placed physics formulae in the middle and was then able to remember them.

The comedian Chella Quint uses what she refers to as her 'slinky of time' for booking gigs, named after the toy invented in the 1940s by Richard James, a naval engineer working on a horsepower meter for battleships. When a spring fell off the table and onto the ground he noticed that it seemed to have a life of its own. His wife came up with the name and the Slinky was born. Decades later, this is how Chella Quint pictures time:

'I see time as a spiral going infinitely out ahead of me and slightly upward into the future, and infinitely down behind me at an angle for the past (my own past and world history as well). The year is basically a circle, with the spiral to the next level at the New Year. I can use my slinky calendar to remember events by compressing the slinky. If I was trying to remember one winter vacation, I can look at all the Decembers in a column (because when I compress the slinky of time, all the months line up at the same point in the circle) till I find the one I want. I have no control over seeing the compressed coil shape and I'm not choosing to do it – I just do.'

Many people insist that these images give them an advantage when it comes to quick-thinking and new experimental evidence seems to back up this contention. In Vancouver psychologist Heather Mann gave people a tricky task – one you can try for yourself.

> Recite the months of the year out loud going backwards in threes, starting from November.

It's not easy, but some if you will find it easier than others. The better performers in Mann's test were the people who had a 'map' of the year in their mind displaying all the months in a clear visual format. This made for much quicker calculation.[40] You might be thinking that proficiency in this particular exercise is not a terribly useful skill and you'd be right. But in everyday life we are constantly required to manipulate time-based information, such as working out exactly how many days there are before a deadline or how much leave we have left, and this is where the time/spacers might have an advantage.

THE SNARC EFFECT

One of the most widely used tests in psychological research involves flashing a up a stream of words one at a time on a computer screen, written in different coloured fonts. The respondent has an apparently simple task: to hit as quickly as possible a key on a computer keyboard which the test setter has designated as corresponding to the colour of the word. So, say, d for red and p for blue. Easy – and not

very interesting to psychologists surely? But this exercise
is surprisingly powerful. Say the word flashed up on the
screen is 'doughnut'. Respondents are told at the beginning
of the test that the meaning of the word is irrelevant. The
only thing that matters is the colour of the word on the
screen. So, the respondent just has to hit d on their keyboard
if the word doughnut is red and p if it is blue. It turns out
that a person suffering from anorexia is significantly slower
than the average in doing this simple exercise. Why? Because
their anxiety at the thought of a 'doughnut' gets in the way
of the task in front of them and slows down their ability
to complete it. When the word is 'room' they can do it just
as fast as anyone else. This kind of test has demonstrated
over many years the way the speed of the response to
something flashing up on a computer screen can be used
to tap in to the way a person thinks, without a person being
able to cheat on the test.

So what can a variation on this test tell us about the
spatial visualisation of time? At the University of Bergen
in Norway, Mark Price asks time/space synaesthetes to draw
a diagram of how they see the layout of the months of the
year. He then sits them at a computer that flashes up the
months randomly on the screen. This time there are no
coloured words. All the respondents are asked to do is press
one designated key for months early in the year and another
key for those later in the year. What Price has found is this:
if a person's mental map of the year has, say, March up in
the left-hand corner then they are quicker at hitting the
designated key for an early month if that key is on the left
of the keyboard. The same person will be slower if the

designated key for an early month is N – a key in bottom right-hand corner of a keyboard. The mental map is never referred to and in theory people should respond at the same speed, whatever the location of the computer key. But they cannot help but picture the mental map, making time faster when the key they have to press happens to fit in with their personal map.[41]

This finding has the rather nice name of the SNARC effect – nothing to do with poetic monsters, but an acronym for Spatial Numerical Association of Response Codes. I had a go myself in the experimental psychology lab at Sussex University run by Jamie Ward. The results were striking. Although the difference in my reaction times was only a matter of milliseconds, the pattern over hundreds of attempts was clear. Every time the designated key on the keyboard for the early months was on the left (for me the 'correct' place on my mental diagram of the year so where I would expect to find January, February, etc.) my reaction time was faster. You might say, 'Well, you knew what you were being tested for.' True. But, even with me, the test was cheat-proof. It all happened so fast there was no way that, even if I'd wanted to, I could have factored out my mind map of the months.

DO WE ALL SEE TIME IN SPACE?

We have heard earlier in this chapter that people are sometimes shy of discussing their ability to see time in space. I hope those reading this book will decide that they should now come out loud and proud as time/space synaesthetes.

It's a useful skill to possess, but only if you allow yourself to use it to its full potential. Why struggle to picture the months in line with an Outlook planner or Filofax diary if you see them in a circle or a three-dimensional spiral? If you work with your brain, rather than against it, aren't you more likely to remember important dates like your mother-in-law's birthday or when to file your tax return? So let it happen. Draw your own time map onto a whiteboard or into a diary and you can improve your memory for time-related events.

Now you might think that this piece of advice only applies to the 20 per cent of people who picture time automatically. However, there is evidence to suggest that we can all use space to code time to some extent. This is not a new idea. In fact the notion is centuries old – John Locke discussed it as early as 1689 and William James described dates as having positions in space back in the nineteenth century. Maybe they were both space/timers. At Ghent University, Wim Gevers found that when a study required people to draw a plan of how they visualised the months of the year, even those to whom it did not come naturally were able to draw something.[42] So I think the ability to picture time is a continuum: at one end are the people who instantly see ladders and slinkys and at the other are the people who have never wondered how they picture the year until pressed to do so.

Here's something else to try.

Draw three circles on a piece of paper representing the past, the present and the future. You can put the circles anywhere on the page, touching or not, and they can be of different sizes if you want them to. There are no right or wrong answers.

While you are doing it, I will tell you about the person who invented this test: Thomas Cottle. He conducted his investigations into time perception in the 1970s using US Navy personnel, who were all accustomed to following instructions.[43] They were told to do the exercise, and so they did. Cottle found that 60 per cent of his respondents drew three separate circles, with the future represented by the largest circle, and the past by the smallest. Most didn't make the circles overlap, leading Cottle to conclude that people (or US sailors anyway) tend to see the past, present and future as discrete time periods.

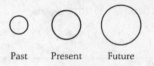

Past   Present   Future

Cottle was not terribly happy with these results. He thought, rather unfairly in my view, that this atomistic view of time was rather childlike. In his opinion it would be logical for the circles to overlap like in a Venn diagram to suggest the connectedness of time and the impact the past has on the present and the present on the future.

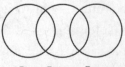

Past   Present Future

Some people make the future circle very large to represent the vast unknown ahead of us. I found myself doing the opposite and making the future smaller than the past or present because, unlike the past, I have no idea what it holds, so for me it feels as though there's less information to fit in there.

To examine further the way we perceive time in relation to our own lifespan and our place in history, Cottle used timelines. Draw a horizontal line and then make four marks on it representing the start of the following points in time – your personal past, your personal future, the historical past and the historical future after your life has ended. Here's mine, but again there are no right or wrong answers – yours might be very different (and never mind whether Thomas Cottle thinks it is childlike).

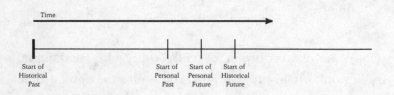

Some people draw the timeline so that their own life takes up the majority of the space – an egocentric perception (within psychology this is not meant as a judgemental term); while others see their life as a short part of a longer line of the world's past and future – a historiocentric perception.

Back in the 1970s Cottle suggested that the historiocentric view suggests an ascriptive orientation – in other words you believe that it's the past that affects your life, rather than your own efforts. He said upper-class people inheriting their wealth would be a good example.[44] An alternative of course is that the historiocentric view is put forward by someone who remembers lessons at school on how short their life is in the context of human history, let alone the earth's history. We were told that if the history of the planet is represented by the distance from your nose to your outstretched fingertip then one stroke of a nail file would wipe away all human history. Your own life within that would be too small to see.

The fact that when pressed we can all represent time graphically, and have a sense of a 'correct' way of viewing it, suggests that to an extent we are all able to associate time with space. Time is difficult to comprehend and difficult to grasp. It's my contention that visualising it in space, even to a limited extent, makes it easier for us to grapple with. We constantly need to think about the past and future, and picturing their positions in relation to our bodies could simplify the concept for us. And as we'll see, this association could even influence the metaphors we use in language. Or is it our language that affects the way we associate time with space?

TIME, SPACE AND LANGUAGE

When laying out either the circles or the timelines, English speakers invariably put the past on the left and the future on the right. You probably didn't even consider laying it out the other way round; it just seems the obvious way to do it, whether or not you usually visualise time in space. Other experiments with random groups of people who have English as their first language have shown that almost everyone presented with flashcards printed with the words past, present and future and asked to lay them out in order on a table will put them side by side in a horizontal line with past on the left, the present in the middle and the future on the right. What is driving this tendency? Is it evidence that most people do visualise time in space even if they're not aware of it?

Rigorous testing has certainly shown that there is a strong association inside the minds of English speakers between the word 'past' and the position 'on the left'. People are not just thinking 'well, if I have to put the past somewhere I'll put it on the left'. The association is stronger than that. Again the SNARC test provides the evidence. Tell people to hit a left-hand key whenever a word associated with the past is flashed on a screen and they will do it faster than if they are told to hit a right-hand key. The past and the left is just *right* somehow.

I've heard researchers suggest that this phenomenon can be explained by the direction in which the hands on a clock move. It's true that the hands on a clock set off from the top of the hour in a rightward direction, so you could argue that this implies the future is off towards the right. But this is a theory that falls apart after 15 seconds, because after quarter past the hour the tip of the hand starts moving towards the left – time is going backwards into the past! Admittedly, by quarter to the hour the steady tick of time is now back on its rightwards trajectory – but only for 30 seconds. As you can see, this is going nowhere. We need a better explanation. More plausible, surely, is that English-speakers *read* from left to right. The very phrase 'left to right' illustrates the point. If you are reading that phrase in

English you read 'left' before you read 'right' – putting left before right in time. Or to put it another way, 'left' is in the past by the time you read 'right'. And here's the clincher perhaps. Arabic and Hebrew are written right to left on the page. Where do Arabic and Hebrew speakers place past, present and future on a left/right spectrum? With the past on the right, the present in the middle and the future on the left – the mirror image of English speakers. This then begs a much bigger question, and one which taps in to the debate that has raged for decades about whether language or thought comes first – does a Hebrew speaker think of the past as being on the right because she writes right/left or does she write right/left *because* she sees the past on the right?

Lera Boroditsky, a psychologist at Stanford University, has done some fascinating work comparing English and Mandarin speakers and the way they discuss time in terms in space.[45] The experience of time should be universal – a moment is with us and then gone; that happens wherever we live and whatever language we speak. But the way we describe that experience differs from language to language. Boroditsky found that speakers of both English and Mandarin use spatial references to horizontal and vertical planes when talking about time. The 'best is ahead' is horizontal, 'let's move that meeting up' is vertical. But she found far more horizontal metaphors in English: we 'put events behind us' and 'look forward to the party at the weekend'. In Mandarin people make more use of vertical metaphors – earlier events are 'up' or 'shang'; later events are 'down' or 'xia'.

Boroditsky stood next to people, pointed to a spot directly

in front of them and asked, 'If this spot here represents today, where would yesterday and tomorrow go?' Unlike tasks using computers, this technique had the advantage of being in 3D. If people saw time as a sash encircling the body, like some of my informants, they were free to point that out. Follow up questions included: if the spot in front of you represents lunch, where do breakfast and dinner go, and if that same spot is September where would August and October go? She found that Mandarin speakers, whether living in Taiwan or California, were eight times more likely than English speakers to lay time out vertically, usually pointing up into the air for earlier events and down for later ones.

There would appear to be an obvious explanation for this difference. Traditionally, Mandarin is read in vertical columns, from right to left. That is changing however. These days it is often set out horizontally across the page from left to right, just like English. Yet the vertical conception of time has persisted and even the people who could *only* read Mandarin horizontally were still seven times more likely to arrange time vertically in space.

At one level expressions such as time 'creeping up on us' or 'flying by', are partly explained by our desire to keep our language fresh and lively. We do not *literally* experience time in this way. Having said that, the language we use to describe time does tell us something important about the essential character of our experience of time. Not least the capriciousness, strangeness and mutability of that experience.

With the exception of the language spoken by the Amondawa tribe in the Amazon where no word for time exists, most other languages constantly make reference to it using words relating to space or physical distance. We talk of long holidays or short meetings, but rarely do we do this the other way round, borrowing the language of time to describe distance. We talk of time speeding up as though it were a physical object in space, like a car, but wouldn't describe a street as four minutes long. But how much does this really tell us about the way we think of time? Do we use time-related phrases because they fit conveniently into our sentence structures or does this tell us something about the way we perceive time? Our experience of time is quirky and unsettling – and we come up with expressions that try to capture that feeling.

Is the way we think about time also influenced by the words we use? Psychologist David Casasanto compared the use of metaphors for 'time as distance' and 'time as amount' in four languages. In English we talk about something taking a long time (implying distance), whereas the Greeks use a phrase meaning physically large to refer to time and Spanish speakers refer to '*mucho tiempo*' meaning much time. Using that handy research method of comparing numbers of hits on Google, Casasanto investigated whether the phrase 'much time' or 'long time' appeared most frequently. It was clear that French and English speakers preferred the use of 'long', the distance metaphor, while Greek and Spanish speakers used 'much', the amount metaphor. But the intriguing part of the study comes next.[46] A group of English and Greek speakers were given a series of tasks presented on a computer screen to investigate whether the way they *spoke* about time also affected the way they *thought* about it. Some tasks involved having to estimate how long it took a line to spread gradually across a screen. Others involved guessing the time taken for a container to fill up with water. Some involved both. The results were clear: the English speakers were distracted by distance and allowed it to affect their estimation of time, while the Greek speakers were distracted by amount. However, Casasanto found that people's loyalty to their language metaphors could be weakened; he was able to train English speakers to think of time in terms of amount instead of distance.

This experiment might seem obscure, but if it really is the case that the language you speak affects the way you conceptualise the relationship between time and space to the extent that it can change the judgements you make about speed, distance, volume and duration, then this is remarkable. This is a fairly new area of study, but you have to wonder what the implications might be. Could the words we use affect our whole attitude towards time?

TIME AND SPACE MIXED UP

Our use of language is not the only evidence demonstrating our association of time with space. In fact there's more than an association. We get time and space mixed up. The father of developmental psychology, Jean Piaget, studied the way children's minds work at different stages of their development. He conducted a study where two trains on parallel tracks move for exactly the same amount of time, but because one is going faster it stops several inches further away than the other. Young children insist that this train must have been moving for longer. Piaget concluded that young children find it hard to distinguish between size as it relates to time and size as it relates to space. Children's brains are of course still developing, but experiments conducted by Lera Boroditsky suggest that as adults we don't find it much easier.[47] We are good at judging distances, but these judgements can skew our estimations of time. So if a series of dots are bunched together as they cross a screen we are likely to think they are moving faster than if they are more spread out, even though they are moving at precisely the same speed. We find it hard to judge time without allowing spatial considerations to come into play.

We are lucky enough to have complex brains that not only compute in many dimensions, but are *conscious* that they are doing so too. Our brains are very clever, but this cleverness can trip itself up. In this case, our brains are *fooled* by their very awareness that time and space are related. Bigger sometimes means faster, but not always. Lions are quicker than mice, but bullets are even faster. In everyday life we constantly make mental calculations involving speed, time and distance – think of catching a ball or crossing a road. So perhaps it's not surprising if they are associated and sometimes confused in the mind. Show children two lights and ask them which was on for longest and they will choose the brighter one. Show them two trains racing and they'll say the biggest one is the fastest. They get the idea of largest, but often apply it to the wrong property, which takes us back to the theory I discussed in the previous chapter: that we might have a neural structure which judges magnitude rather than time in particular. As adults we make fewer of these mistakes, but the relics of this space/time crossover are still there.

There is one rather mysterious element to all this: the way we think about time and space is not symmetrical. If you show people a line of three light bulbs, switch them on one at a time and ask people to guess the time between each light coming on, the more spread out the lights are physically, the longer people will say the duration between lights was. This is known as the 'kappa effect'. It's similar to the study with the series of dots moving across the screen, and also works the other way round. If you switch the lights on in turn and ask people to estimate the *distance* between them, the faster you turn on the lights, the closer they will say they are. This is called the 'tau effect'. Just as we know a lion is big so it probably runs fast, we find it hard to ignore what we know about speed and distance and assume that faster probably means nearer. But Boroditsky and Casasanto have shown that the relationship between space and time is imbalanced. We think about time in terms of space *more* than we think about space in terms of time. This brings us back round to the language and the lack of phrases such as 'a four-minute-long street'.[48]

Rhesus monkeys do things differently, displaying a symmetrical interference of time with space and space with time that shows they are just as likely to think about space in terms of time as they are time in terms of space.[49] Is this because they don't have language or because their senses differ? We know rhesus monkeys can't learn to throw a ball like a human being. Does this mean they haven't learned about the way that force, time and distance interact; that the harder you throw a ball, the further it goes, but the longer it takes to land? Time and distance (or space) seem to have a unique kind of association in the human mind. Perhaps those fanciful mental images of decades and days of the week that I discussed at the start of the chapter mean much more than we think. They could even allow us to do something extraordinary – to represent mental time in space so that we can time-travel mentally in a way that no other animal can. At will, we can imagine next week or think about when we were seven and then jump back again. This remarkable ability is something I'll discuss in more detail in Chapter Five. In my view it's the fact that we can picture time at all that allows us to think about future events, and also to imagine impossible events. At will, I can picture – and any moment now so will you – a small mouse flying up to the moon on a toothbrush on New Year's Eve next year, dodging fireworks as it flies. No one knows what a monkey might be able to conjure up in its mind, but is its imagination stunted by an inability to spatialise time?

WHEN IS WEDNESDAY'S MEETING?

The way we associate time and space is not just a theoretical matter; it affects us on a daily basis in the real world too. We all use space to think about time, some of us, as we've seen, more elaborately than others. There's one simple question that highlights the differences in the way individuals view time in space, one which separates us into two groups.

> Next Wednesday's meeting has had to be moved forward by two days. What day is the meeting happening now?

There are two possible, equally correct answers to this question, yet I've been surprised at the number of people who before answering insist that they know this is something they always get wrong. You can see them struggling not to give their first, intuitive answer. What they really mean is that they have encountered situations where another person feels differently and have then blamed themselves for somehow misunderstanding time. Although there's no right or wrong answer, whether you answered Monday or Friday tells you far more about the way you personally see time passing than you would ever guess. If your first instinct, regardless of how you think others might answer the question, is that the meeting is on Monday, then it is time that is moving, like a constant conveyor belt where the future comes towards you. You are using the time-moving metaphor.

Time-moving metaphor

If you believe the meeting to be on Friday, then you have a sense that *you* are actively moving along a time-line towards the future – the ego-moving metaphor.

So either you stay still while the future comes towards you or *you* move along towards the future. It's the

Ego-moving metaphor

difference between thinking that we're fast approaching Christmas or that Christmas is coming up fast. Did you pass the deadline or did the deadline pass you?

THE RIVER OF TIME

In captivity in Gaza, the BBC journalist Alan Johnston had a strong image of time as a flowing river or a sea. He even visualised the water deliberately to help him deal with the hours he was forced to spend with nothing to do but think. Mental imagery is a strategy psychologists sometimes teach people to help them to cope with difficult situations such as living with chronic pain, but Alan developed this coping mechanism of his own accord. With no one to talk to and no idea when he would be released, it's not surprising that he had what he modestly refers to as 'bleak moments'. But Alan had the same determination shared by the psychiatrist Victor Frankl during his time in Nazi concentration camps – that his captors could control everything else, but not his thoughts. He resolved to take charge of his own mental state.

Alan took comfort in the notion of a river, because while the repetitious, empty days and nights suggested endless circularity, the flow of a river meant that he was always moving forwards, and that one day things would be different:

'I used to imagine myself as a boatman on this river of time. I knew the river would eventually get some-where. Either I'd die of old age in that place or some moment of freedom would come. But regardless, something *would* happen; I would not always be living like that. Even if I reached old age there, somehow the river of time would reach a point where this thing would stop. As a boatman on the river of time I needed to keep my eyes on the horizon and not to endlessly watch the bank passing by because that would feel slower. I used to think that I needed to keep steering the boat into the calmer waters, by which I mean a calmer state of mind. When I was miserable and unhappy and dwelling on dark things, the river had choppier waters and flowed less swiftly, so the thing was to steer yourself into the calmer, faster-flowing water of the river of time which would eventually, inexorably bear you to an end of that situation.'

As a free man, Alan had read about Ernest Shackleton's journey in a tiny open lifeboat after his ship the *Endurance* had become trapped near Elephant Island in the ice floe that eventually wrecked it. If Shackleton could take that little boat across a dangerous ocean and even survive a hurricane, then Alan could survive his own journey locked in this room in Gaza. 'I had to make a mental journey across the endless wastes of sea and time. Just as Shackleton was aiming for a tiny speck of land far, far beyond his horizon, I was aiming for a moment in time far, far beyond the time horizon – the moment when I'd be free.'

He imagined that his vessel on this 'sea of time' was a raft of mental planks. Each plank represented something positive about his situation – his health seemed to be okay, Gaza was a place where deals were often made and he hadn't been tortured.

'I imagined lashing those mental planks together to make my raft on the sea of time. I had a powerful mental picture of me and this vessel. When you're feeling low, you think how am I going to get myself into the brighter side of my mind again? You argue yourself out and every time you use the argument it gets slightly blunter. You hear yourself saying where's the freaking deal been for the last three months? Then instead of going through the arguments all over again, you envisage that raft slowly pulling you through. You see yourself on the sea of time and that image is quite calming. It's symbolic. Sometimes there's a mental storm and the planks will be scattered in the sea and you have to swim around and pick them up. You're thinking how good each plank is and you lash them back together again and off you go. That was a process that went on many, many times. I hadn't thought about it till now but my ways of coping were time-related.'

Alan's image of himself as constantly moving on towards the future is one shared by many people. Regardless of what he did, he knew that time would carry on taking him onwards and he would reach the future. This is an example of the ego-moving metaphor (again, ego is not being used in a judgemental way in this phrase). It contrasts with the time-moving metaphor I described, where you remain motionless while the future comes towards you. Does the day of an exam get nearer to you or do you get nearer to the day of the exam? This is why the question I posed earlier about moving Wednesday's meeting is so revealing. It gives you a clue as to which of these two perspectives you personally ascribe to. I remind you again that there's no right or wrong answer here, no better or worse way of viewing time.

MAKING TIME GO BACKWARDS

The incessant forward movement of time has long fascin-ated writers, who have attempted to explain it through tales of time-travel or even through the reversal of the direction of time. Lewis Carroll's rarely discussed novel *The Story of Sylvie and Bruno* features a brother and sister who sometimes appear as children, sometimes as tiny fairies. When they encounter the Outlandish Watch, which makes time go backwards, people they meet take pieces of lamb out of their mouths and return them to a joint of meat, which rotates slowly over a spit until it becomes raw again and eventually the fire dies down to a single flame and is extinguished. The accompanying now-raw potatoes are handed back to the gardener for burial and of course every conversation becomes nonsensical. In Martin Amis's novel *Time's Arrow*, the Nazi doctor heals his Jewish patients of their terrible wounds rather than inflicting them. In *Counter-Clock World*, Philip K. Dick takes it a step further, with old people knocking on the insides of their coffins begging the funeral directors to unearth them from the ground in the cemetery. Once rescued, these 'old-borns' live their lives backwards, gradually getting younger. If they've written a book earlier in life the state ensures that every copy is destroyed at the precise moment it was published. Time continues to rewind until they are children and eventually babies, whom women then volunteer to carry. Their pregnancy bumps get smaller and smaller until the day of conception when the women have an overwhelming desire to find someone to have sex with. This unconceives the old person. They are no more and it is as though they had never existed.

These are only stories, but new research shows that with the clever use of a mirror, making time appear to go backwards is easier than these writers could have guessed. This experiment even took its creators, the psychologists Daniel Casasanto and Robert Bottini, by surprise.[50] They took Dutch speakers and performed the usual test of getting them to press keys on either the left or right of the keyboard when different time words flashed up on the screen. As you'd expect for speakers of a language written from left to right, they were fastest when they had to press the left key for the words in the past. Then they did the same experiment but with the words written on the screen in mirror-image, forcing them to read from right to left, and everything changed. The direction of the flow of time reversed in their minds and suddenly they were faster when the early words were associated with a right-hand key. This may sound simple to you now, but I can assure you that it really was an extraordinary discovery and I can see why the authors were so surprised. This is a young area of study, and at the moment we don't know what the practical implications of this finding might be, but its significance lies in the suggestion that the direction in which we read affects not only our thinking, but the way we visualise time itself.

Your answer to the Wednesday meeting question might seem intuitive, but this too can be manipulated. Your response can even depend on where you're standing. Lera Boroditsky is particularly imaginative when it comes to devising experiments that illustrate just how easy it is for us to switch our perception of time depending on what we're doing at a particular moment.[51] She asked people the same Wednesday meeting question, but some of them were on the Caltrain from San Francisco to San Jose, others were at the airport and the third set of people, in true psychological research tradition, were in the researcher's own department at Stanford University queuing for lunch. This is the same psychology department where Philip Zimbardo conducted his famous prison experiment, turning the basement into a makeshift jail (more on that in Chapter Five, but I've seen the so-called solitary confinement cells they used in that study, and they really are just cupboards only a filing cabinet deep). I'm pleased to say Boroditsky's study involved no such cruelty, but nonetheless led to findings which are significant.

In response to the Wednesday meeting question, people

who are waiting – either to board their plane or at the back of the lunch queue – are more likely to give the answer Monday (the time-moving metaphor). They are waiting for time to come to them, to allow them to start their journey or eat their lunch. The people already getting onto the train or about to get off it and those disembarking from a plane were more likely to say Friday (the ego-moving metaphor). They had actively begun their journey and already felt that they were moving forwards, rather than being forced to wait for time to come to them.

It is through experiments like this that you can begin to see how our feelings about a situation affect our perception of time and the way we visualise it. In one experiment people were asked to imagine an event they were either dreading, like an operation, or anticipating happily, like a wedding.[52] If they were looking forward to it, they were more likely to see themselves as moving towards it, but if they were dreading it, it moved towards them. Emotions and time are undoubtedly connected as we've already seen, and space is one of the factors that enables this connection. This last experiment makes me wonder whether the time-versus ego-moving metaphor might be a reflection of a person's general optimism or pessimism about life. Are those who see themselves marching ever onward into the future more optimistic? This is another piece of research I'd like to add to my wish list of future experiments.

MELLOW MONDAY AND FURIOUS FRIDAY

This next study might seem rather peculiar, but bear with

it. We've already established that it's not uncommon to ground time in physical metaphors associated with space or distance, and that your emotional thoughts about an event can make a difference, but this study takes things a step further. The rather lovely title 'Mellow Monday and Furious Friday' might give you some idea of where this is going.[53] The experimenters took the familiar Wednesday meeting question, but this time they specifically asked people about the emotions they were feeling. They found that if people felt angry they were more likely to see themselves moving through time (the Friday answer and the one associated with looking forward to an event) than to see time moving towards them (the Monday answer, more associated with dread). Then something really extraordinary happened. If the days were arranged on the computer screen in such a way that people were induced to give Friday as the answer, this then changed the way they felt. They then rated themselves as feeling angrier, for no apparent reason.

We know that negative emotions such as disgust, fear and dread are pull-away emotions that encourage you to withdraw. But anger is different; instead of feeling the urge to run away, you want to attack. When you're furious you might slam the door as you march out of the room, but this is usually to avoid saying or doing something you might regret later. It takes an effort of will to leave the source of your anger rather than confront it. Anger draws you towards your target. The authors of the 'Mellow Monday' study believe that when we think about ourselves moving towards the future, we associate this with moving towards something, which we link with feeling angry. It's an interesting idea, but I'm not sure enough evidence is there yet to back up this chain of thoughts and events. They even suggest that if this association between emotions and the way you see time is as powerful as it appears, then one method of deliberately calming yourself when you feel angry is to think of time moving towards you, rather than you to it. Not easy to do in practice of course, but it might be easier the other way round. If you are dreading the exams that are soon approaching you, you could try imagining yourself in control striding purposefully towards them – if you can bear to.

In this chapter we've seen how our perception of time is influenced by the language we speak, the direction in which we read words on a page, our moods and even our position on a journey. Our sense of time in space is profound, and it enables us to time-travel mentally. On a whim we can decide to imagine life when we've retired or to think back to our first day at primary school. This ability is called chronesthesia and it's something that I'll be discussing in more detail in the subsequent chapters. Incredibly, mental time-travel can even be expressed in our bodies. If people are asked to envisage a typical day four years ago while standing blindfolded, they begin to lean back a few millimetres without even realising it. When they imagine a typical day four years into the future they lean forwards.[54] No one in this study guessed what it was about, suggesting that their swaying bodies were not influenced by the experimenters. What this tells us is that time and space are embodied.

Even if we don't visualise centuries of kings and queens

laid out on decorating tables, to an extent we do all experience time within space, whether it's through the way our bodies move when conceptualising time, the location of the past and the future in relation to our bodies, or our sense that time is like a river. Language provides an insight into the way we perceive time, but also seems to shape that thinking, illustrating that once again we *create* our own perception of time in our minds. Time constantly surprises us and confuses us. We can't write it down. We can't see it. We can't capture it. So the ability to picture time even to a limited extent helps us to manipulate it in our minds and paves the way for mental time-travel. Next, we're going back in time – to the past.

WHY TIME SPEEDS UP
AS YOU GET OLDER

HAVE A LOOK at this list of events. Without looking them up, can you name the year and the month in which they took place?

John Lennon is shot dead

Margaret Thatcher becomes British Prime Minister

Chernobyl nuclear power plant explodes

Michael Jackson dies

The film *Jurassic Park* is released in the USA

Argentina invades the Falkland Islands

Morgan Tsvangirai is sworn in as Prime Minister of Zimbabwe

Hurricane Katrina strikes New Orleans

Indira Gandhi is assassinated

A car bomb explodes next to Harrods in London

The first cases of swine flu hit Mexico

The Berlin Wall comes down

Prince William marries Kate Middleton

An IRA bomb explodes at the Grand Hotel in Brighton

Barack Obama is inaugurated as President of the USA

Princess Diana dies

Bombs explode on the London Underground

Saddam Hussein is executed

33 miners become trapped in a mine in Chile

The first Harry Potter book is published

The answers are at the end of this chapter. The years are easier to guess than the months; nonetheless the chances are that you will only get some of them correct. This is normal, but the nature of any mistakes you make illustrates something extraordinary about the way the mind organises the past. I'll be referring back to this list many times during this chapter. Most of us think we're bad at remembering names, but in fact Japanese research has found that we're more likely to remember the names involved in a news story than the date.[55] Luckily we don't get tested on dates very often, so it's something we don't tend to realise. You probably thought that some of the events happened more recently than they did and that they feel so familiar that you're surprised that they took place so many years ago. Perhaps this even stirs in you uneasy feelings that time has somehow slipped by without you noticing. Does this seem to be happening more often? As you get older is time speeding up? You think you saw someone a few months ago, only to realise that it was in fact last year. You expect your friend's child to be a toddler, only to discover they've been at school for years now. Their growing up is a repeating reminder that time moves on.

The sensation that time speeds up as you get older is one of the biggest mysteries of the experience of time. In this chapter I'll show why the key to time speeding up lies in our perception of the past. And memory is to provide the explanation for some of time's other curiosities too. I'll begin by examining how autobiographical memory works and along the way I'll be telling the tales of some people who have gone to extreme lengths both to record and test their own memories of the most everyday events. I'll be covering the different theories about why time speeds up, concluding with my own, which I call the 'Holiday Paradox'. This can also explain why it is that a good holiday appears to whizz by and yet afterwards you feel as though you've been away for a long time, and why the days go slowly, yet the years so fast, if you're bringing up small children.

We know that time has an impact on memory, but it is also memory that creates and shapes our experience of time. Our perception of the past moulds our experience of time in the present to a greater degree than we might realise. It is memory that creates the peculiar, elastic properties of time. It not only gives us the ability to conjure up a past experience at will, but to reflect on those thoughts through autonoetic consciousness – the sense that we have of ourselves as existing across time – allowing us to re-experience a situation mentally *and* to step outside those memories to consider their accuracy.

AUTOBIOGRAPHICAL MEMORY

In July 1969 the British tennis player Ann Jones took part in what should have been the most memorable match of her life. She was playing in the Wimbledon Ladies' Final against Billie Jean King, who'd already won the title three times, meaning Ann wasn't the favourite. Nevertheless the match went to three sets, and after Billie Jean King served a double fault on match point, Ann Jones became the champion. After competing in 13 Wimbledon tournaments she had finally achieved her dream. Princess Anne presented her with the trophy, she held it up for everyone to see, the crowded applauded and the photographers captured the moment. You would assume this would be the one tennis match planted firmly in her memory forever; yet speaking on Ladies' Final day 40 years later, she confessed that she

barely remembers it. 'People expect me to remember everything about the final, and always ask me about it, but over
time it all blurs. I remember the semi-final better.' She says
she does recall the feeling of winning and that it was everything she'd hoped it would be, but ask her the score and
she has no idea. The BBC sent her a video of the match
and although her children and grandchildren love it, she's
never watched it herself. Her experience shows that even
memories of unique, personally momentous events can
fade. Most of what we do is forgotten. When we talk about
the study of memory, really it should be the study of forgetting. Every day we experience hundreds of moments that
we simply forget.

The study of memory has been a major area of research
within the field of psychology, yet compared with the number
of studies conducted on topics such as short-term memory
or semantic memory, the ability to recollect personal experiences has been relatively neglected. Autobiographical
memory can be divided into two types: episodic memory,
which consists of specific personal experiences, for example
arriving on your first day at a new school; and semantic
memory, which consists of the knowledge we have about
our lives and the world, and would include the *facts* about
the school you went to – the town it was in and how many
pupils attended.

To make sense of our own memories we rely on our
understanding of time. Whenever we tell the narrative of
our own lives it is natural to link events, to put them on a
timeline and to explain how one led to another. Back in
1885 the philosopher Jean-Marie Guyau said that just as

cities build on top of earlier civilisations – 'the living city is built on top of sleeping cities' – so the present covers up the past in our minds, constantly building on top of it. But just as archaeologists can uncover Roman mosaic floors under modern buildings, if you look carefully many of the ruins of memory remain. At the time we tend to see the choices we make in our lives as relatively independent of the times we live in, but looking back we can explain retrospectively how our story fits into its place in social history. If you ask someone why they are choosing to embark on parenthood in their mid-thirties, at least 10 years later than their parents might have done, their explanation will tend to relate to their personal circumstances rather than the trends of the time. So they might say they hadn't met a partner in their twenties, or wanted to finish their education, travel or get their career under way, not that social or political factors have pushed them in a certain direction. But ask the generation above and they will say that they had children in their early twenties because that was what you did then. These patterns are hard to see in our own lives, partly because we want to believe our personal choices are just that, personal, and not swayed by the cohort to which we belong by an accident of history.

TOTAL RECALL

Having a conversation with Gordon Bell is slightly unnerving. While you stand opposite him you know that the small, black device on his chest is photographing you every 20 seconds, creating pictures that he intends to preserve forever in his extensive slideshow of his every waking moment since 1988. He calls it 'Total Recall'. It's not just photos. He saves everything – every bank statement, every email, every text, every webpage he visits, every answerphone message (including all the times his wife has told him to turn off the recording), every TV programme he watches and every page of a book he reads (he has even employed a patient assistant whose job it is to scan these pages). In theory you could choose any date from the past 24 years and relive his life on that day, seeing everything he saw and reading everything he read. He describes his methods with enthusiasm, clearly fascinated by the technology that has allowed him to do it and the complex systems he uses to file everything digitally. But you can't help but feel slightly sad, wondering whether anyone will care. He is creating an extraordinary record of one man's life, but a life where a lot of energy is devoted to finding methods of keeping an extraordinary record of a life. Will anyone look at all this information after his death? Maybe they will. Maybe in a few centuries' time he will be the Samuel Pepys of his day, although I get the feeling he includes less gossip in his diary. Also, he's not alone. He has a competitor.

When the Reverend Robert Shields died in 2007 he left behind 91 cardboard boxes containing a type-written diary spanning 25 years which chronicled his life, minute by minute. It's 30 times longer than Pepys' diary. It makes Robert B. Sothern's decades of daily bodily measurements seem as brief as a simple medical note. The reverend's commitment was such that he woke up every two hours during the night, every night, in order to write down his dreams. During the day he would sit in his thermal under-wear in his office in the back porch of his home in Dayton, Washington, surrounded by six electric typewriters arranged in a horseshoe. Every page of typing consists of a list of times accompanied by a description of exactly what he was doing, from shaving to opening junk mail. His father had won the title of world speed-typing champion for his ability to type the Gettysburg Address repeatedly at 222 words a minute. We don't know whether Robert inherited his father's gift for typing, but even if he did it still took him four hours a day to write his diary. He bequeathed it to Washington State University on condition that no one read it until 2057. At approximately 37.5 million words, it *might* be the longest diary in the world, but until it's unveiled in 2057 no one will know because not even an exact word count is permitted. In the few extracts which have been made public, the events are mundane. He changes light bulbs. He watches *Murder, She Wrote* starring Angela Lansbury (he notes that 'The action in *Murder* is quick, fast and condensed.'). He eats macaroni cheese. He trips up in the street while carrying home some leftovers after supper with friends. He records the precise number of sheets of papers he uses every time he goes to the loo. He finds copious different ways of describing urination: 'I let out my water tank,' and 'I hosed down the ceramic facility until it surged with foam.'[56] Mundane, but curiously gripping. Believing that the histo-rians of the future will want to study both his diary and his DNA, he has taped samples of his nose hair to one of the pages.

This is the ultimate autobiographical record, one with which a human memory could never compete. He and Gordon Bell are both determined not to let the fallibility of memory destroy the record of their lives. This seems to be a trend, with numerous life-bloggers noting down every detail online. Gordon insists his e-memory is different and more comprehensive than the life blogs, but his motivation is still curious. He works for Microsoft and travels the world giving talks on the project and the technology he employs, yet is hesitant to discuss practical applications. You can see ways in which this kind of electronic memory could be extremely valuable – for someone with memory problems as a result of brain damage, for example. Yet his aim, he says, is only to demonstrate that it is possible to record a life. He shows us films he's made by editing together the photos taken at 20-second intervals. They remind me of those flicker books we made as children where you get a notebook and draw a man diving off a diving board, leaning over slightly more on each page, until he's submerged and all you're left with is a splash and then a ripple. In Gordon Bell's film streets jump by and food disappears bit by bit. It's stilted, yet super-fast. But these films are just for illustration. He stores his whole life, he says, but rarely plays it back. His colleague asked him whether this could be described as a memory that is WORN – standing for *Written Once, Read Never*. He doubts that. He thinks someone will watch it all.

The beauty of these attempts by the reverend and Gordon Bell is that they remove the big limitation of autobiographical memory: its selectivity. But that might also be the

problem. Perhaps one day we'll each have a digital library of our own lives and the ability to pick any day and watch it again. I'd choose one date at random and look at the pictures for that same date every year, hoping to get some sense of how the world had changed or how I'd changed. You could relive the best parties, experience a day in your first job or pretend it's Christmas every day. But would you get round to actually doing it when it comes at the expense of creating new memories, or at least watching films where you don't know the ending? How many couples have watched their wedding video more than once or twice? If this kind of technology meant you remembered everything, it could have a profound effect on your perception of time because, as we shall see, it is through the lens of autobiographical memory that we gain a sense of time gone by. We rely on these memories of the past, to a greater extent than we might realise, to judge the speed of time as it passes in the present.

WHEN TIME SPEEDS UP

Misdating the year of the Princess Diana's death, or the fall of the Berlin wall, is just one feature of life speeding up as you get older. While a week with nothing planned stretches ahead of you if you're 11 and it's the summer holidays, take a week off work as an adult in the hope of redecorating your house and before you are even halfway through the painting, the week will have rushed past. Anyone over the age of 30 will tell you time is accelerating and that any markers of time, whether it's Sunday nights or

Christmas, appear to come round sooner every week or year. It came as a shock to many when the news was released in 2001 that the two 10-year-old boys infamous for murdering the British two-year-old Jamie Bulger were now grown up and that plans were being made for their adult lives. It wasn't so much the surprise that they would be free so soon, but that they could possibly be adults. With Jamie Bulger forever frozen in time at the age of two, the age gap makes it all the more unsettling. Part of this is our revulsion at the crime, but it is also a reminder that whether we like it or not time has moved on.

The feeling that time speeds up as you get older is very common in adults, and one that makes no sense to children. I can remember the irritation I felt as a little girl when adults would marvel at how I'd grown. It seemed like a stupidly obvious remark. Now, although I try very hard not to say it out loud when children can hear me, I can see what a startling marker of time their growth provides. What intrigues me about the sensation of time speeding up is that it's something we often discuss, but to which we never seem to become accustomed.

The first explanation most people give involves straight-forward mathematics. A year feels faster at the age of 40 because it's only one fortieth of your life, whereas at the age of eight a year forms a far more significant proportion. This is known as proportionality theory and has been supported by many over the years, including the writer Vladimir Nabokov. It tends to be credited to the nineteenth-century French philosopher Paul Janet, who wrote, 'Let anyone remember his last eight or ten school years; it is the space of a century. Compare with them the last eight or ten years of life: it is the space of an hour.'[57]

The workings of autobiographical memory surely do come into the explanation for time speeding up as we get older, but not necessarily by way of the proportionality theory proposed by Janet. In fact even back in 1884, the philosopher and psychologist William James wrote that this theory of proportionality gives more of a description of the phenomenon than an explanation, and I have to agree. 'The same space of time seems shorter as we grow older – that is, the days, the months, and the years do so; whether the hours do so is doubtful, and the minutes and seconds to all appearance remain about the same.' The problem with the proportionality theory is that it fails to account for the way we experience time at any one moment. We don't judge one day in the context of our whole lives. If we did, then for a 40-year-old every single day should flash by because it is less than one fourteen-thousandth of the life they've had so far. It should be fleeting and inconsequential, yet if you have nothing to do or an enforced wait at an airport for example, a day at 40 can still feel long and boring and surely longer than a fun day at the seaside packed with adventure for a child. If this doesn't convince you, and the proportionality theory is one that many insist feels intuitively correct, then think back over the past week. If you are an adult then in lifetime terms this single week is completely insignificant, yet at this moment it feels alive in the mind and relevant. The events which took place might not matter at all in 10 years' time, but they will have an impact on this week and even next month perhaps. Janet's theory is neat as a description, but it won't suffice as an explanation because we simply don't consider the context of our whole lives when judging how fast recent months or even the past year has passed. It ignores attention and emotion, which as we've already seen can have a considerable impact on time perception. The theory fails to explain all the different situations where time can warp. I've already mentioned enforced waiting, and there is the strange impact that a holiday can have on the experience of time. When people return home they often comment that they feel as though they've been away ages, yet if the proportionality theory held, and we considered that fortnight as a proportion of a lifetime, then it should feel tiny, almost unmemorable.

We should be relieved at the lack of evidence supporting proportionality theory, because the consequences could be depressing. If the proportionality theory is correct then a 20-year-old who eventually lives to the age of 80 would already have lived half of their subjective life. These figures come from a formula devised by Robert Lemlich in 1975.[58] When he asked people of different ages how fast they felt time was moving, he found their answer was predicted by his formula, according to the theory of proportionality. However, later research has found that it doesn't quite work. According to Lemlich's theory, a 60-year-old should feel as though time is going twice as fast as it did when they were 15, but if you ask the 60-year-olds how fast they feel time is moving now compared with when they were 15, the average answer they give is that it is only moving 1.58 times faster.[59]

You'll have noticed already the problem with all this: it is

all based on a person's subjective experience of time, and
the subjective is never easy to measure. Although it's common
to say that time feels as though it's speeding up as you get
older, it is surprisingly difficult to demonstrate. If you ask
people to look back on their lives, they will invariably tell
you that time feels as though it's going faster than when they
were young, but this relies on their memory of how time felt
all those years ago. When today's 75-year-olds were 25, no
one asked them how fast the years were passing, which means
that we have to rely on comparing today's young people with
today's older people. This opens up the possibility that it's
the tempo of life in general that's changed, rather than their
personal perception of time as they age. Today, both younger
and older adults claim that time passes quickly. In a Dutch
study more than 1,500 people were asked how fast they felt
the previous, week, month and year had gone. More than
three-quarters answered 'fast' or 'very fast', regardless of
age.[60] Perhaps it is the case that life goes slowly as a child,
when you have little control over what you do, and once you
reach adulthood it goes fast for almost everyone. But there
is one question you can ask that does highlight a change
with age and which concerns the speed of the previous
decade. The greater a person's age, the more rapidly they
will say the previous decade elapsed. So perhaps days, months
and years don't speed up, but there's something special about
the decades.

If the proportionality theory is considered more of a
description than an explanation, how can we make sense
of the decades speeding up? The answer is up for debate,
but the main theories involve the workings of

autobiographical memory. Which brings us back to the list of events at the start of this chapter.

LIFE THROUGH A TELESCOPE

Most people say they feel daunted by the task of trying to date public events from recent history but then go on to get many of them right. Have a look at any events you misdated. This is where it gets interesting, as mistakes can tell us a lot about the workings of the mind. Did you tend to think that events happened more recently or longer ago than they actually did? There are revealing patterns in these dating errors that give us a window onto the problem of time speeding up. The chances are that for events that happened at least 10 years ago, such as the explosion at Chernobyl nuclear power plant or Princess Diana's death, you thought some of them happened more recently than they did. This common mistake is known as forward tele-scoping. It is as though time has been compressed and – as if looking through a telescope – things seem closer than they really are. The opposite is called backward or reverse telescoping, also known as time expansion. This is when you guess that events happened *longer ago* than they really did. This is rare for distant events, but not uncommon for recent weeks. You might think you saw a friend three weeks ago, but in fact it was only a fortnight ago.

Forward telescoping is one of the factors that can contribute to the sensation of life speeding up, and I'll come back to why it might do that later in this chapter. First I want to look in more detail at the phenomenon of telescoping. The most straightforward explanation for it is called the clarity of memory hypothesis, proposed by the psychologist Norman Bradburn in 1987. This is the simple idea that because we know that memories fade over time, we use the clarity of a memory as a guide to its recency. So if a memory seems unclear we assume it happened longer ago.

When it comes to dating news stories, we might assume that the more we know about an event, the more able we will be to name the date it happened. It seems not. When Susan Crawley and Linda Pring from Goldsmiths College, University of London, gave people of different ages a list of events very similar to the one I've given you, but longer, the following events were those this British sample found it easiest to date, giving both the correct month and year: Margaret Thatcher becoming Prime Minister, Margaret Thatcher resigning, the shooting of John Lennon, the invasion of the Falklands, the Grand Hotel Brighton bomb, the Chernobyl disaster, the assassination of the Israeli Prime Minister Rabin, the Dunblane Massacre, the Baltic Ferry disaster, the Lockerbie plane crash and the hurricane that hit southern England. Surprisingly the amount a person knew about an event *only* made a difference to the accuracy of dating it if it happened before they were born. For events that happened within our lifetime we don't appear to use knowledge to date them.[61] Instead we rely on memory. The exception, of course, is when we know so little about an event that we've not even heard of it. Then we tend to assume that it must have happened long, long ago because otherwise we would remember it. The events used in these studies vary depending on where they are conducted. In the list of a dozen news story events used in research in New Zealand, only two rang any bells for me, although I wish I had known about the story of Shrek, the extra-woolly sheep who had not been shorn for many years and apparently became a media star after he was discovered during a muster in Central Otago.

The incident people in the British study found hardest to date was the hijacking of an Ethiopian Airlines plane, which crashed in the Indian Ocean in 1996. In every age group most people had no idea when it happened and assumed it must have been a very long time ago. Compared with the other events on the list that took place that year, such as the opening of the Channel Tunnel and the Baltic Ferry disaster, I can't say I remember it myself. But when I looked it up I discovered the most extraordinary story, one that wouldn't be easy to forget. On a flight from Addis Ababa to Nairobi, three young men charged the cockpit, shouting into the intercom that they had just been released from prison in Ethiopia and wanted political asylum because their anti-government views left them in danger in their own country. They claimed to have a bomb, which later turned out to be a bottle of drink, and had chosen this particular aeroplane on the basis of an article they'd read in the in-flight magazine, which said it could fly all the way to Australia without refuelling. This, they thought, would suit them perfectly. What they didn't realise was that because the plane was only on a six-hour stopover flight, the fuel-tank wasn't full. The pilots pleaded with them, but the hijackers were convinced they were lying and insisted they steer the plane

in the direction of Australia. Knowing they couldn't make
it that far, the pilots continued along the coast in the hope
of making an emergency landing at the airport in the
Comoros Islands just as the fuel ran out. This hijacking was
unusual because for four hours the hijackers allowed life in
the cabin to proceed as though little had happened. The
passengers were aware of the hijackers and even made plans
to overpower them once they'd landed, but they had no idea
of the arguments occurring in the cockpit. They continued
to eat, read and, rather surprisingly considering the circumstances, to sleep.

As they approached the Comoros Islands, just as the
pilot predicted, the fuel started to run low. He knew this
was his one and only opportunity to land the plane safely,
so he began dropping altitude. As soon as they noticed
what was happening, the hijackers wrestled for the controls
and the disruption caused the pilot to miss the runway and
ditch the plane in shallow water just off the island he was
aiming for. This is one of the few occasions in airline history
that a plane this large has landed on water. Despite the
diagrams in the safety instructions where planes float on
the surface of the water while passengers calmly remove
their high heels and slide down the emergency chutes ready
to blow their whistles to attract attention, large planes rarely
float. They sink. And even landing in shallow water, this
crash was to prove fatal for many of the passengers. One
side of the plane hit a coral reef, causing it to break up.
Scuba divers on holiday rushed to the rescue alongside
locals, but still 123 of the 175 passengers and crew died in
the crash. Today the incident is told as a cautionary tale in

safety training for airline crews as many of the passengers who survived the impact and managed to get their life jackets on made the fatal error of inflating them *before* escaping from the cabin – the life-jackets pushed them up towards the roof of the plane, which was by now full of water and the people drowned.

The pilot, Leul Abate, survived the ordeal and was given an award for his bravery. One of those who lost his life was the cameraman and photojournalist Mohammed Amin, famous for the pictures he took of the Ethiopian famine of 1984. Coincidentally, by the time of the hijacking, he had become the publisher of the Ethiopian Airlines' in-flight magazine, the same magazine used by the hijackers to decide which plane to seize in their attempt to escape from the country.

As I said before, this really is an extraordinary story and the chances are that I'll remember it the next time I hear it mentioned, and probably so will you. But because knowledge doesn't improve the dating of events which happen in our lifetimes, we still won't necessarily remember how long ago it took place. The event isn't tethered to a particular date for us. The fact that so many people don't remember this hijacking highlights one of the difficulties of studying autobiographical memory and working out the frequency of the telescoping of time. You can't test somebody on their ability to estimate the date of the Ethiopian hijacking if they've never heard of it. Studying short-term memory is easy; you can give a whole group of people the same list of words to memorise, test them under different conditions and score them on their accuracy. But while news events may seem to be universal, they aren't. If you've never heard of an event like the Ethiopian hijacking you will never remember the date, however fantastic your memory. An alternative solution is to test people's autobiographical memory for personal events instead, but this brings two new problems: not only are everyone's memories different, but they are hard to verify. I remember going with my grandfather to an air show where a motorcyclist attempted to jump over a line of double-decker buses. It was the climax of the day and hundreds of us stood watching. It looked like an impossible task. Would he manage it? Surely he'd crash. He started far, far away from the buses, roared up the ramp and took off. Then, as he hovered in the air, the crowd gasped as it became clear he wouldn't make it. He fell onto the buses, glancing off the edge of a roof and landing on the grass. The ambulance staff rushed to rescue him, but it was too late. He was lifted onto a stretcher and an orange blanket was pulled over his head. I remember it well. My grandfather tried to stop us from looking and took us to find the car. Or maybe it didn't happen like this at all. My sister tells me it wasn't an air show, but an agricultural show; it was our elderly neighbour who took us, not our grandfather; and the motorcyclist wasn't killed; he did fall, but he only injured his leg. My sister is four years older than me and she's probably right, but our differing stories illustrate just how tricky it is to assess autobiographical memory and the part it plays in time perception. If every memory needs verifying, how are we to work out who's good at it and who's not?

TAKE TWO ITEMS A DAY FOR FIVE YEARS

Psychologists have avoided some of these problems by asking people where they were on a certain date and then verifying the answers via diaries or relatives. But one researcher tried something a little more extreme, a technique of which I suspect Gordon Bell might approve. Back in 1972 Marigold Linton's idea was to have someone write down everything that happened to them, however insignificant it might seem. Then in years to come the accuracy of each autobiographical memory and its place in time could be tested. Looking around for a suitable guinea pig for such a study, Marigold Linton sought an individual who was easily accessible, reliable and, crucially, prepared to take part in a daily study with a five-year commitment. Following in the footsteps of many a scientist in the past, she decided there was only one subject who would do for the job – herself. Her conscientious nature was never in doubt – as a member of the Cahuilla-Cupeno tribe of Native Americans, she was the first ever person from a Californian reservation to attend college. On opening her first report card at college and seeing she had straight As, she was so surprised that she tried to return it to the office, convinced it must belong to someone else. Nevertheless she found researching her own memory far more challenging than she could ever have predicted.[62]

She called the study 'Take-two-items-a-day-for-five-years', although really that name only covered the first part of the study. In fact every evening for 10 years she sat down in her home in Salt Lake City, took a fresh white filing card measuring six inches by four and typed three lines describing an event that had happened to her that day. She considered each experience for a moment, then rated its confusability, emotionality, importance, datability, the likelihood that she would discuss it with others and whether it belonged to a sequence (e.g. one lecture in a course of twelve). On the reverse of the card she wrote the date and then shuffled it in among the other cards from that month. The first day of each month was a testing day (in both senses of the word as it transpired) where she would select two of the previous months' cards at random and guess which occurrence happened first and the date, all the while timing herself with a stopwatch. The idea was to assess her ability at a very particular type of time perception – setting events in their place in time.

The Marigold Linton approach – noting down daily memories for testing on a later occasion – has been attempted with larger groups of people too. The problem is that this test can never provide a measure of autobiographical memory as a whole, only of the events a person selects. Inevitably these will be the most outstanding and therefore the best-encoded in their memory. So the selectivity of memory affects what can be tested later on – you're unlikely to write down that you dropped a letter on the ground while you were trying to post it, but because you don't record the incident, you can't be tested on your memory of it either. Yet these diary studies do begin to give an insight into which events we remember, how we order them and how those autobiographical memories begin to build up a sense within us both of time and of our life history.

Linton replaced each index card after testing, which meant that by chance some events would come up more frequently than others. It was these events which she became best at dating, demonstrating the more a memory is discussed or contemplated, the more likely you are not only to recall the event, but to remember the date. The events of 9/11 would be an extreme example of this, the one incident whose date we can never forget, because it is not only mentioned frequently, but is named after the date it took place. When it comes to personal memories, we might expect to remember best the moments where things went wrong, but Linton found the opposite. Again this is probably down to mental rehearsal. We might worry for a while about the time we embarrassed ourselves in front of a roomful of people, but we don't look back on photos of the day in the same way we do with our eighteenth birthdays (although with more and more moments – good and bad – documented on social-networking sites, this might change in the future).

This is known as the fading effect bias – the somewhat counter-intuitive idea that while the negative loses its sting in the memory as time passes, the positive doesn't lose its joy. The theoretical explanation is that discussing an incident from the past has a different impact on that memory, depending on whether it was good or bad. So every time you talk about the good old days you relive that memory and the warm feelings that accompanied it, but with the exception of the extreme case of post-traumatic stress disorder, unhappy events gradually lose their power the more we talk about them. This allows us to cope and move on.[63] We even remember successes as having taken place more recently and embarrassments as more distant in the past, as though time warps to protect our self-esteem.[64]

In a series of studies using diaries, the psychologist John Skowronski asked students to keep a daily record of one occurrence that was unlikely to happen more than once in a term. He found the results fascinating and confesses that one of the attractions of this work is the amount people reveal about themselves anonymously, even when they're asked to be discreet. Two months later each student was given two of their events at random and had to guess which came first, as well as the day and the date. Women did slightly better than men, but still usually got the date wrong. Some hypothesise that women score higher because they tend to have the calendar-keeping job in a family, but these were young students, so I would question whether that would apply. Perhaps they initiated and arranged more social events, which led to them knowing the date. Not surprisingly students were more likely to get the date wrong the longer ago the event took place; for every week that passed they were out by one more day. The day was easier to get right than that date, so people might have known it was a Tuesday, but found it hard to assess which Tuesday it might have been. Weekends had boundaries of their own and were clearly delineated, so people might feel certain an event happened either on a Monday or Tuesday, but would never say it might have been a Sunday or a Monday.[65]

It seems that an event's place in time comes low down in our memory's priorities. After keeping daily records for six years, a Dutch psychologist, Willem Wagenaar, found that the what, who and where of an event were well-remembered, but the *when* simply didn't feature as much.[66] The reason I'm interested in these studies is that they can tell us more about the phenomenon of telescoping – when you think events happened more recently than they actually did – and crucially, whether this contributes to the sensation that time speeds up as we get older.

The results of these studies reveal that telescoping can occur with personal memories, just as it does with the recall of news stories. It makes no difference whether the event was pleasant or unpleasant, and Skowronski found once again that if an event was poorly remembered we tend to assume it happened longer ago than it really did. At first sight this seems to make sense and fits in with the clarity of memory hypothesis. We know memories fade, so if someone can't remember an event very well you would expect them to assume it must have happened long ago. These theories, sometimes known as 'trace strength' theories, date back to the nineteenth century; the stronger a trace we have for a memory, the more recent we assume it to be. But in fact this theory doesn't hold. Although we do assume poorly remembered occasions happened in the distant past, when there's a personal experience we remember well, we are often bang on with the right date, especially if it was during the last four months.

Nevertheless we do get some dates wrong, and it can matter – and not just in pub quizzes. This phenomenon can even affect public policy. Surveys involving questions about events inform everything from policies on anti-social behaviour to insurance premiums. When a pollster phones

up it is standard practice to put a time-frame on the events
they are discussing. If they want to monitor your use of
local leisure facilities, they are not interested in knowing
that you once went to the local swimming pool in 1999;
they want to know about your behaviour in the last 12
months. This helps them to ensure that their findings are
up to date. If the local council commissions research to
assess the impact of their policy of increasing community
policing in the streets in your area, they don't want you to
remember an incident where you felt threatened five years
previously. They need you to recall only incidents from the
last year. The problem is that people often get things wrong.
When I took part in a local crime survey, I wanted to tell
them about the time when two 10-year-old boys on their
way home from school pointed a plastic gun at me and
shouted, 'We're going to shoot you up your arse!' (And no,
I haven't made that up, so you can see why I'd want to tell
them about it.) Then I realised it had happened a least a
couple of years previously, but our tendency to forward
telescope the salient events means that – without wanting
to mislead – I could have easily given the wrong information. If everyone else did this too, crime figures would
appear higher than they really are.

Something similar can happen with insurance questions.
We are asked for a history of car accidents within the past
three years. We even sign a form saying we've told the truth,
but since dating events is difficult, whatever our intentions,
we might not have done. Car accidents are events that are
both unusual and alarming, so you could expect them to
stick in the memory; but since we already know that

negative memories lose their power over time, it shouldn't surprise us to learn that people often forget about accidents altogether. By checking records against drivers' recollections one study found that as many as a quarter of road accidents are forgotten.[67] Add this to the impact of telescoping and many a survey could have inaccurate results that then go on to influence policy. The provision of GP services, for example, relies not only on records of actual visits, but on asking people how often they have visited in the past three years. If everyone includes a few extra visits by mistake, this could skew the data considerably. When 200 students at the University of Alberta were asked how many times they had seen a doctor during the previous two months, many of them included appointments which happened far longer ago.[68] And one of the reasons (and of course there are others) why people fail to visit the dentist as regularly as recommended is that each trip feels fairly recent.

The chief dentist at the Winter Olympics in Canada in 2010, Dr Chris Zed, told me that considering Olympians are elite athletes who take very good care of their bodies, they have surprisingly poor teeth. Part of the reason that 75 dentists were present at the games was to deal with the inevitable accidents – anyone who missed that second Wu-Tang on the ski cross could easily smash their jaw and need a dentist – but they were also there to take the opportunity to get inside the mouths of the athletes. Their itinerant schedules make regular check-ups difficult and because they are hardened to withstand pain, they continue training with the kinds of abscesses that would leave the rest of us howling. It wouldn't surprise those 75 dentists if the next time some of those competitors see a dentist is during the 2014 winter Olympics in Russia. These skiers and skaters will probably intend to have a check-up between now and then, but they're busy people and the time will pass quickly. It is only in 2014, reclined in the dentist's chair, that they will realise that not only have they not seen a dentist since Vancouver in 2010, but also that four years have passed. Only then will the marker in time provided by the games alert them to the years which have passed since their last check-up. But it's these kinds of markers that can help solve the problem of skewed surveys.

> Think back to the last two months and count the number of friends you have spent an evening with.

The chances are you will include friends whom it feels as though you've seen recently, when in fact you saw them several months ago. It's easily done. But it's also easily avoided; this method improves the accuracy of surveys and our own time estimates by changing the wording of the question to include an explicit landmark in time. So instead of asking, 'How many times in the last year have you been to see your doctor?' you ask, 'How many times since New Year's Day have you been to see your doctor?' The landmark, in this case New Year's Day, gives people a firm anchor in time, making it easier for them to calculate which events came before it and which after. When reconstructing memories we make a minimal cognitive effort, but if we're asked for an actual date we are forced to compare a memory with other landmarks and are more likely to get the answer right.

TIME-STAMPING THE PAST

Bob Petrella is a middle-aged American television producer who remembers everything. He remembers every conversation he's ever had and everywhere he's ever visited. When he lost his mobile phone he wasn't concerned about losing his contact numbers: he knows them all off by heart. This is because he is one of only 20 people in the world who have been diagnosed with a newly discovered condition called hyperthymesia, or Superior Autobiographical Memory. It was discovered accidently by an American neuroscientist called James McGaugh, who had spent his career studying memory. In the year 2000 he was contacted by a woman who wanted to tell him about the problem she had. He was accustomed to this, and patiently explained that his department at the University of California Irvine researched memory problems, but wasn't a treatment facility. But hers wasn't exactly a memory problem – more the opposite – as she told him she doesn't forget anything. Intrigued, he agreed to meet her and soon realised she was telling the truth. She could remember everything. Since then another 19 people have been found to have this rare ability, including Bob Petrella. Never mind naming the dates of a list of famous events like the one at the start of this chapter; he can do it the other way round too. Give him a random date and Bob will be able to tell you the event that happened that day. When he was at school he found exams easy and didn't really understand why other people felt the need to revise. He seemed to know everything, except the fact that his mind was rather exceptional. James McGaugh is now studying the brains and genetic make-up of Bob and the other nine individuals who share his abilities, to try to discover how they do it. He has already seen structural differences in both their grey and white matter and hopes that in the long-term their remarkable abilities might shed light on memory processes that can in turn help those who suffer with memory problems.

Bob remembers the date of every football match he has ever seen. He excels at getting dates *exactly* right. Although the rest of us get plenty of dates wrong, we can identify approximately 10 per cent of dates correctly. Sometimes we can work them out through reconstruction, linking them to other memories from that month or year. Occasionally we simply know an exact date is correct, without having to trawl through our memories to prove it. The mystery is how we do this. One theory is that on occasion we create a memory that comes with some kind of time-tag attached. This time-stamp tells us when it happened and could explain this sporadic accuracy. What it doesn't explain is why the other 90 per cent of memories should fail to have this time-stamp.

If I'd asked you to put the list of news stories at the beginning of this chapter into chronological sequence instead of guessing the actual months and years they took place, you would have found the task much easier. However, people with damage to the parts of the brain associated with remembering new facts find the sequencing of past events very difficult, illustrating once again how crucial memory is to our perception of time. The neurologist Antonio Damasio has found that people with amnesia also lose these time-tags, finding it impossible to distinguish which events happened in which decade. This deficit presents a real problem; if we can't create and store memories, we are unable to get a sense of the chronology of our own lifetimes and our place in the world. When Damasio asked healthy people to lay out personal and public events from the past along a time line, they were on average two years out. When people with amnesia due to damage to the basal forebrain did the same task, they were wrong by an average of just over five years. But, and this is where it gets interesting, people whose amnesia was caused by damage to a different area, the temporal lobe, remembered the events less well, but could still time-stamp. This suggests that the details of the memory itself and the time-stamping of that memory rely on different processes. Damasio has found that this corresponds with his observations of patients; those with damage to the basal forebrain can still learn new facts, but might get those facts in the wrong order.

When you're trying to gauge when certain events happened, you might be able to work out one time-frame, but not another. You could have no idea of the year, but be certain that it happened on a Saturday. Time does not have a linear hierarchy in the same way that other types of memory do. With faces, for example, if you are given a picture of an actor you may or may not remember his name, but you will most likely remember his job. This is because his job exists at a higher level in your memory. People don't say, 'That's Ethan Hawke, but I can't remember what he does,' they say, 'That's an actor, but I can't remember his name.' With time it's different. Take, for

example, the death of Princess Diana. This has become one
of those flashbulb memories, like the assassination of John
F. Kennedy, where everyone can tell you exactly what they
were doing when they heard the news. People don't necessarily know the exact date, but they will probably remember
the day of the week – because Diana died late in the middle
of a Saturday night most people didn't find out until they
woke up on Sunday morning, and for many people Sundays
are distinctively different from other day, making them
more memorable. If it had happened on a weekday, they
would find it harder to remember which weekday it might
have been. You are also better at remembering news events
that happened on a day that held a personal meaning for
you, so the events you never forget are where the personal
and the public cross over. If Michael Jackson died on your
thirtieth birthday, and people were discussing his death at
your party and requesting his songs, the chances are you
will forever remember the date of his death.

To sum up, you are most likely to remember the timing
of an event if it was distinctive, vivid, personally involving
and is a tale you have recounted many times since.

EVERYTHING SHOOK

On the morning of Friday, 31 January 1986 a woman was
shopping at a mall in Mentor, Ohio. At 11.48 a.m. she
was wondering what to buy. Everything seemed normal.
But one minute later nothing seemed normal at all. Goods
were falling off the shelves, the clothes racks were swaying
and the whole place seemed to shudder. She didn't

understand what was going on. People began rushing for the nearest exit and just as she started to move she felt something smack on her head. Putting her hand up to her face she felt blood and as soon as she saw the ceiling tile that had hit her on the head, she knew what this was – an earthquake.

The rumours soon began – people had been killed; houses were wrecked. In fact there were no fatalities and no one's home was destroyed. The earthquake was relatively mild – only 4.96 on the Richter scale. Some 15 people were treated for anxiety or for the effects of the cold; a little girl had stitches after she was cut by a broken window; and doctors were soon able to deal with the bleeding head-wound of the woman who had been out shopping. In earthquake terms this wasn't serious, but for thousands of people this did become a day that was in a small way unique. They might have been one of the dozens who phoned the local geological society; one of the hundreds evacuated from the nearby nuclear power station; one of those who noticed that the water in their local well had changed colour; or Betty, the school bus driver who told the local paper, *The Spokesman*, that she'd lived through tornadoes and floods, but had never seen anything like this; or the Mayor of the town of Sharon, who watched his staff flee as a four-foot crack appeared in the wall of the municipal building.

There is a long history of psychologists grabbing unusual opportunities to study situations they could never recreate in an artificial experiment. There was the eminent expert in visual perception, Richard Gregory, who was reading the newspaper one day back in 1958 and discovered that surgeons had restored the sight of a man who had been blind for 50 years. Here was the perfect chance for him to study whether eyesight automatically lets you watch and comprehend the world or whether the brain has to spend many years learning to make sense of the input from the eyes. He filled his car with the instruments needed for the assessments he wanted to carry out and drove to the hospital to find the man, known in the literature as S.B. The resulting case study became world-famous (answer: for full vision we do need to *learn* to see). More recently there was Barbara Frederickson, who happened to have tested the psychological resilience of a group of students several months before 9/11. She found herself with a unique opportunity to explore how a shocking episode affected levels of optimism and how this linked to a person's underlying levels of resilience (a surprising answer: the most resilient people actually felt *more* optimistic after 9/11 than before).[69]

It was a psychologist called William Friedman who spotted the potential of the earthquake in Mentor, Ohio, as a way of studying time perception. As I've already mentioned, the problem with testing people on the timing of news events is that – like the Ethiopian plane hijacking – not everyone has heard of every event and even when they do it can be hours, days or months afterwards. But all the people in the earthquake were instantly aware of it.

Nine months after the earthquake, Friedman sent a questionnaire to every employee of the nearby Oberlin College, to ask them to guess the time, date, day, month and year it happened. Most people could guess the time to within an hour, but they found it very hard to name the day of the week.[70] Once again this gives us some insight into how we reconstruct time in the past using fragments of information, clues which are absent on weekdays. Friedman found that even four-year-olds can tell him the time of day of an event, but it's not before they are about six that abstract concepts like months start to make sense to them. On occasions, we use indirect elements of a situation to attempt to come up with a date – what was the weather like, was it dark? Or we link it to something that has a firm date in our minds already – did it happen near Christmas? You might work out the year of the Falklands War by remembering that Margaret Thatcher was in power or that you were at school or college. Researchers Alex Fradera and Jamie Ward (the same Jamie Ward who works on synaesthesia) found that when people were encouraged to map out timelines of their own life events on paper and then add the news events to the same timeline, they did better than when they guessed the dates without reference to their personal lives, regardless of whether they considered the news story to be striking. This is a strategy you can adopt deliberately when you need to work out the date of a news event – think of as many links as possible to the details of your personal life at the time.

There are other techniques for dating events and reducing the impact of telescoping that I'll come back to in Chapter Six, but here's just one more for now. When the psychologist John Groeger asked people to recall any car accidents they might have had, they remembered more when told to work from the past to present than vice versa. The famous psychologist and false-memory expert Elisabeth Loftus found that you can improve accuracy even more if you ask people to think of occurrences within a longer time period first and then to narrow it down to a smaller one. So if you really want to know how many times you've been to the doctor in the past six months, think of a personal landmark from roughly a year ago, think your way from back then through to now and then answer the question again, just concentrating on the last six months.

A THOUSAND DAYS

To work out the timing of an incident in your personal life, two months is the cut-off date you need to remember. It shows up in studies again and again. If an event happened more than two months ago, the chances are it happened a bit longer ago than you think. So if you guess that it happened six months ago, add a month on and you'll probably be closer. If it you guess it was eight years ago, it was probably nine.

We find public events harder to date, and it's possible they are processed and stored differently by the brain. There is a tipping point that crops up time and again in research – 1,000 days, or roughly three years. Three years seems to be the time-frame we assess best. When Chuck Berry told me the story of his glider crash, he guessed that it had happened three years before, but wasn't certain. I looked for a report of his accident to check the date. It was two weeks short of three years. He had got it right. Now of course these are average results, so it won't happen with every single event, but if you look back again at that list of news events, you might find that the ones that happened three years ago are the ones where you were correct. If you think a news event happened more than three years ago, then add some time on, adding more the longer ago you think it took place. But if you suspect it was less than three years ago, but more than three months ago, then you are likely to have *under*estimated how long ago it was, reverse telescoping the time. A strange fact that I will come back to in more detail in Chapter Six.

So all of this tells us something about how we place events in time and how to get better at guessing dates, but what about the bigger question we're trying to answer – why time speeds up as you get older. Is the phenomenon of telescoping robust enough to explain this sensation that time is distorted? First, there's a mathematical issue. In many studies of telescoping the participants are told that the dates they are going to be asked to recall all fall within a certain time-frame, such as the previous six years. Mathematically it is inevitable that the errors people make will skew towards the middle of this time period. They know they can't give a date longer ago than six years, which encourages them to give more recent dates. This could account for some of the apparent telescoping.

Age also makes a difference. In Susan Crawley and Linda Pring's classic study, 'When Did Mrs Thatcher Resign?', which opens this chapter, the list of events was given to people in three different age groups. The 18- to 21-year-olds were let off the most distant events, but it was the 35–50 age group who did best; they did better than the over-sixties. When they did make mistakes, the middle-aged group forward telescoped in the way we might expect if time speeds up as you get older. But the over-sixties did something very different. They made more mistakes, but not through forward telescoping. They had a tendency to date events as happening longer ago than they did, something psychologists had thought only occurred with events from recent weeks, not distant events like Margaret Thatcher's resignation or John Lennon's death.[71] While the older people were very good (and in fact better than the other age groups) at dating some news stories, such as the storming of the Iranian embassy in London by the SAS and the Challenger Space Shuttle disaster, they thought that the massacre in the English town of Hungerford had happened six and a half years earlier than it did. Were the middle-aged people particularly interested in the news? Or had the over-sixties become so accustomed to time flying and getting dates wrong that they had deliberately over-compensated by dating them too far in the past?

Perhaps it's lucky that we're more often required to remember names than dates because researchers in this area often comment that people dislike taking part in these studies. The task sounds straightforward, but some go as far as to say they find it painful to learn that they don't know when a famous event took place. Why should we find this so troubling? Skowronski believes that the pain is caused by the close relationship between our judgements of time and our self-image. When there has been a big change in self-image the passage of time is easier to judge; any new parent would find it very easy to divide events into pre- and post-baby. And it's this close intertwining of identity and time perception that causes us discomfort when we cannot work out timings. We feel that we own the news events that have happened during our lifetimes; that they are almost a part of us. With personal memories the feeling is even stronger, and so when we don't know when something happened we lose the sense that we are in control. Willem Wagenaar, the Dutch psychologist who kept a diary for six years, described the experience of putting his own auto-biographical memory to the test as not just boring, but downright unpleasant.

Marigold Linton did not enjoy her self-experimentation either. After choosing herself as the ultimate reliable partic-ipant, she soon became disappointed with her efforts. As the years went by and there were more memories to include, the testing on the first day of each month began to take many hours; she would test herself on as many as 215 different events. She wrote that she had 'looked forward to working with a completely tractable subject, one who would come on time and would be motivated and consistent. This impression is simply not true. I am frequently intractable, resentful, and distractible, especially as a long test day drags on.' Sometimes she forgot to take part in the experiment at all.

Dull though these sorts of studies might be for the partic-ipant, they are crucial to our understanding of the experience of the years passing us by. However, telescoping does not provide us with the ultimate explanation for why life feels as though it speeds up as you get older. The studies of both personal and public events show that telescoping does not occur as consistently as we might expect. If the theory held, then the older you are, the more you should forward telescope time, but in fact it is the middle-aged who forward telescope time the most. Telescoping undoubtedly occurs some of the time and research on this topic can tell us plenty about how to date events more accurately in our own lives, but – along with the theory of proportionality – it still fails to provide a full explanation for why time speeds up as you get older.

THE REMINISCENCE BUMP

Think back on your life and try to recall a couple of experiences that made you especially happy and a couple which made you sad or afraid. How old were you when these events took place? There is a good chance that some if not all of them took place when you were between the ages of 15 and 25. Psychologists have found that people have a preponderance of memories stemming from this period of our lives. This is known as the 'reminiscence bump'. And this bump could be the key to life speeding up as you get older.

The reminiscence bump involves not only the recall of incidents; we even remember more scenes from the films we saw and the books we read in our late teens and early twenties. If you look back at the list of news events, many of those you dated correctly are likely to have occurred during your own bump. The bump can be broken down even further – the big news events that we remember best tend to have happened earlier in the bump, while our most memorable *personal* experiences are in the second half. It's such a robust finding that it can even help you guess a person's age. Ask them to name someone famous called 'John' from the past and the chances are that they'll choose a John who was well-known when they themselves were in their late teens. From this you can work out their rough age today. In a study which did just this in 1999, people in their fifties were most likely to name John F. Kennedy (US President, 1961–63), while those in their thirties selected John Major (British Prime Minister, 1990–97). For the name 'Richard' the thirty-somethings picked the British TV presenter Richard Madeley, the 40-year-olds chose the pop singer Cliff Richard and older people either stuck with Cliff Richard or chose the rock-and-roll legend Little Richard. Some chose Richard III. He wasn't, of course, alive during their teens, but maybe they saw or studied the Shakespeare play while they were at school.[72]

The key to the reminiscence bump is novelty. The reason we remember our youth so well is that it is a period where we have more new experiences than in our thirties or forties. It's a time for firsts – first sexual relationships, first jobs, first travel without parents, first experience of living away from home, the first time we get much real choice over the way we spend our days. Novelty has such a strong impact on memory that even within the bump we remember more from the start of each new experience. In a study of adults' memories from their first year at college, 41 per cent of memories were from the very first week, the week with the greatest number of new events. That said, novelty does not tell the whole story – childhood is packed with fresh experiences and yet we don't remember that period with as much clarity. We know that the brain goes through a special period of development in late adolescence and early adulthood so one theory, not yet proven, is that the brain might be so efficient at this time of life that it lays down the strongest memories.

My favourite explanation involves identity. We saw how the close intertwining of memory and identity can cause discomfort when our memories let us down. This same connection could shed light on the reminiscence bump. During late adolescence and our early twenties most of us are working out who we are and who we want to be. The Leeds University psychologist Martin Conway, who conducted the study on the famous Johns and Richards, proposes that during this period of identity development we lay down particularly vivid memories, which then remain accessible in order to help us maintain the identity we have created. If this were true, then you might expect people who find themselves going through some kind of major transformation of identity later in life to experience a second reminiscence bump in order to consolidate their new identity. This is exactly what Conway found when he studied the memories of Bangladeshi people who had lived through the struggle for independence from Pakistan and begun new lives in the 1970s.[73]

These three theories that account for the reminiscence bump – brain development, identity searching and novel experiences – are a powerful combination. It's a phenomenon that has been cleverly exploited by TV producers, who worked out some time ago now that we love to wax nostalgic about our teens. Nostalgia is an intriguing emotion. We consider it to be a warm, positive feeling, yet it includes tones of loss and perhaps the longing for a happier past. It has bittersweet elements, so much so that it used to be discouraged, and was even once considered to be a psychiatric disorder. The term nostalgia was coined in 1688 by a doctor called Johannes Hofer to describe the worrying behaviour he observed in Swiss mercenaries who were far from home. They would weep, refuse to eat and in extreme cases even attempt suicide. Over the subsequent two centuries various bizarre physical causes of nostalgia were proposed, including the driving of blood to the brain by changes in atmospheric pressure and the clanging of cowbells in the Alps damaging brain cells and the eardrum. By 1938 nostalgia had been branded an 'immigrant psychosis' and four particular populations were thought to be at risk: soldiers, seamen, immigrants and children in their first year at boarding school. Yet by the end of the twentieth century there had been a sea change and nostalgia became the warm, fuzzy feeling in which we now like to revel.

The ability to time-travel mentally back into the past serves a serious function when it comes to identity. It helps us to cement our individuality and to search for meaning in a life we know to be finite. By examining the past, we can make the future where we cease to exist feel more distant. Nostalgia also performs a social role, reinforcing our connections with other people. When we share so many memories how can we feel alone in the world? It can make the present feel more tolerable by improving our self-esteem.

Strangely we even look forward to nostalgia, deliberately becoming part of events in order to be able to look back on them with everyone else in the future. We want to create memories so that we can say we were there, whether it was Live Aid in 1985 or the 2012 London Olympics.

Nostalgia can even bring comfort in the most hopeless of situations. Victor Frankl admits that when he was in Auschwitz he found consolation in nostalgia. He would deliberately imagine his past life in great detail – taking the bus back to his home, walking up to the front door, getting out his keys, unlocking the door and turning the lights on in his apartment. The memory of these small actions could bring him to tears, but still he felt it occupied his mind and somehow lessened his pain.

The period about which we are most apt to reminisce and feel nostalgic is that within the reminiscence bump. It has even been suggested that the bump could be the answer to the mystery of time speeding up as we get older. If memories from between the ages of 15 and 25 are somehow extra accessible in order to construct and confirm your identity, then it makes sense that this plethora of vivid memories could make youth seem long and therefore slower, while adult life with fewer stand-out moments appears to go faster. This feeling is compounded by the lack of markers in time during middle age. When you are young, you might move home every year or two. It is easy to remember which years you spent at university, for example, and where you went after that, but as you get older and become more settled you move about and change jobs less frequently, allowing the years to merge into one another.

This does provide a partial explanation for the sensation that time moves faster as you get older, but still it cannot provide a total explanation for time speeding up because it applies to such a specific time-frame. It can't explain why time feels faster in your sixties than in your thirties. The reminiscence effect contributes, but to get closer to the heart of the acceleration of time we need to go back to the idea of novelty and its antithesis, monotony.

REMEMBERING MOMENTS, NOT DAYS

The reason Marigold Linton persevered with her unhappy project studying her memories across several years was that she wanted to discover whether William James was correct when he wrote in 1890, in *The Principles of Psychology*, that 'the foreshortening of the years as we grow older is due to the monotony of memory's content, and the consequent simplification of the backward-glancing view'. Referring to time, he said 'Emptiness, monotony, familiarity, make it shrivel up.' James' view fits in with the much more recent ideas about so-called 'memory effects', which are in truth more about forgetting than memory.

Monotony is the culprit; and judging by the contents of Marigold's index cards, the highlights of her days did mainly involve endless cups of coffee and games of tennis. She confessed that the monotony of her own life had surprised even her. Cards would read, 'I had coffee with Jeff', or 'At 4.30 we completed the Xeroxing of the final copy of the statistics book.' To be fair it wasn't all dull. Events which had seemed unimportant at the time took on more significance as the months went by. One day she meets a 'shy scholar'. Nothing is made of this, but later she starts seeing more of him and eventually they marry.[74] Marigold's surprise at her apparently boring life is a perfect illustration of memory effects. We forget the events that we repeat often, while new events stand out and give rise to stronger memories. Try the following experiment:

> Try to remember everything you have done in the past fortnight and don't look in any diaries, emails or paperwork for help. How many events can you think of?

In the fortnight before I first wrote this paragraph I remembered interviewing five or maybe six different people for radio programmes, going to a hen night at a Greek restaurant in London, seeing the film *In the Loop*, and a man on rollerblades almost crashing into me on the pavement and then shouting at me because I flinched. The last two weeks feel as though they've been very busy, but to be honest this is all that stands out for me. Re-reading this paragraph several months later is like an experiment in itself. I would never have guessed that these events were related in time and although I remember the film and the rollerblader, I've no idea what the interviews might have been about. If I was prompted, I'd probably recognise the research findings and remember a few of the details, but it has no anchor in time. As the Italian poet Cesar Pavese said, 'We don't remember days; we remember moments.'

The average person can remember between six and nine events when they're given this task. Yet if you think back to the last time you went abroad, the chances are you will remember far more than nine incidents, particularly if you stayed in several different places. Several months ago I spent twelve days in the USA for work, travelling to seven different cities in three different time zones. I could easily list 30 different memories, from running round a frozen lake in Madison, Wisconsin, to leaning over a wall in Chicago to look at the river which had been dyed bright green for St Patrick's Day, to not believing the bossy sat-nav woman when, while trying to get to our hotel, we pulled into an Ikea car park in a bleak industrial estate just off the freeway and she confidently announced, 'You have reached your destination.' We thought she was wrong, but unluckily for us the motel was right there. I could write pages about that trip, yet it happened far longer ago than that the other fortnight I described. If memory were to fade uniformly with time it should be less clear. Instead novelty stands out.

I mentioned the temporal schemata we develop as we grow up, which give us a sense of what the months of the year and the passing seasons mean. They also give us ideas about how fast time moves and how many events typically fit into a certain time period. We learn to gauge how long events take and then in turn to judge the passage of time by the number of events that have occurred. When the same events are repeated, for example when we go to work, we tend to think little time has passed, and then get brought up short when presented with a definite marker in time like an anniversary. So if we become accustomed to a large number of memories fitting into that decade of the reminiscence bump, then when there are fewer new memories in our thirties and forties, we will feel as though less time has passed and be surprised when we find out that another year has flashed by.

I believe monotony and variety are crucial to explaining many of the mysteries of time. Time drags when you are ill, and you yearn for the hours and days to pass so that you can feel better again. Yet, when you look back, the hours and hours of feeling ill barely feature in your memory. You remember that you *were* ill, but with little novelty a week spent ill at home feels like a lost week in terms of memories. The exact opposite happens on a good holiday. It flies by at the time, yet feels long in retrospect. This brings me to the Holiday Paradox, an effect that can finally explain all these tricks of time.

THE HOLIDAY PARADOX

Every day after breakfast Hans Castorp prepared himself for the morning ahead. He lay down on the reclining chair in his private loggia, took two rugs made from camel hair and deftly wrapped them around himself, one from left to right, the other from right to left, turning himself into a perfect parcel where only his head and shoulders were exposed to the cold mountain air. His reclining seat was made from dark red wood. There was a head-roll at the top and flat cushions covered the length of the chair so as to support his body from head to foot. It was always placed at an identical angle, perfectly lined up with the chairs of the other residents, each in their own loggia. It was quite the most comfortable chair he had ever known. As he looked out over the mountains, Hans knew he was now ready to start the day, a day he would spend resting. Again. This was the moment he loved; the moment when he was able to consider the expanse of time ahead of him – with nothing to do.

Castorp is the young German hero of Thomas Mann's *The Magic Mountain*, a book which pre-dated and seems to have anticipated much of the research on the perception of time. Castorp travels to a Swiss sanatorium to visit his cousin, intending to stay for three weeks, but doesn't leave for seven years. In his first week there is a lot to take in. He learns the routines and meets the other guests. But he soon notices that his strange, empty life appears to be warping time. The others warn him that a week 'above', as they refer to life on the mountain, is not the same as a week back home. Spending so many hours sitting still on his loggia, he wonders whether time passes more slowly when you're not physically moving – remember those studies I mentioned of people getting on and off trains in California. Even the structure of the novel echoes the quirks of time, with the first five chapters describing his long seven years at the sanatorium in minute detail, and then as soon as he leaves time begins to accelerate with the remaining six years squashed into just two chapters.

Thomas Mann believed that novelty somehow refreshes our sense of time and that as soon as we leave our usual routines to visit somewhere new we change time's tempo. This might imply that the solution to creating a life that feels long is to travel constantly, but Mann warns that this new pace of life lasts for only six to eight days before the freshness begins to wear off. The one consolation is that the sense of novelty kicks in again when you first get home and can last a few days or, Mann says, for only 24 hours for those with 'low vitality'.

There is no doubt that Mann was correct in his observation that vacations do curious things to our perception of time. A good holiday passes disappointingly fast. Compared with the months of anticipation and hard work saving up the money to go, the actual time spent away is short. Take a holiday that is a week long. After the first couple of days settling in, you have just two or three days' holiday before you find yourself in the run-up to leaving and already you're calculating when you will need to set off for the airport. It was over in a flash. Conversely, you get home and something strange happens. You look back on your holiday and it feels as though you were away for some time. Could it really have been only a week? You are left with two simultaneous yet contradictory experiences of time. The holiday felt fast while you were there, but afterwards it feels as though you were away for ages. The longer the trip, the stronger the sensation that something is not quite right. This is the Holiday Paradox. Once again William James summed it up for us, 'In general time filled with varied and interesting experiences seems short in passing, but long as we look back. On the other hand, a tract of time empty of experiences seems long in passing, but in retrospect short.' Holidays are the perfect example of the former, while the latter is illustrated by illness, or life on the magic mountain, or in a far more extreme situation like the one in which the psychiatrist Victor Frankl found himself. As well as controlling his own mind, as I mentioned in the last chapter,

Frankl was determined to make use of his time imprisoned in Nazi concentration camps to study the human mind in general. One of the observations he made was that although the days passed slowly, the months rushed by. 'In camp, a small time unit, a day, for example, filled with hourly tortures and fatigue, appeared endless. A larger time unit, perhaps a week, seemed to pass very quickly. My comrades agreed when I said that in camp a day lasted longer than a week.'[75] Frankl's experience fits in with what we already know about the impact of new memories on time perception. The days were very similar to one another. Once people had become accustomed to the routines, and even to the extent of the daily horrors they were living and witnessing, in a curious way they were left with few new memories to create. Frankl himself likened this to the elongation of time described by Mann on the Swiss mountain. Life at the sanatorium relied on strict schedules with meals and rest-cures providing strong, regular markers in time.

Memories and markers in time are two key elements of the way we experience time. Holidays provide the perfect conditions for time to pass quickly – disruption to a daily routine and the removal of cues to the hours passing, combined with a host of new sights and sounds to absorb the attention. The days appear to fly by. When you get home the other key element comes into play – memory. The reason you feel as though you've been away for ages is that so many new things have happened that you have far more new memories than in a normal week, warping your standard mental measurement of time. It is my contention that the Holiday Paradox is caused by the fact that we

view time in our minds in two very different ways – prospectively and retrospectively. Usually these two perspectives
match up, but it is in all the circumstances where we remark
on the strangeness of time that they don't.

If you think back to the studies where people estimated
the time passing while listening to a busy piece of music,
or after getting cold diving, these two methods of mentally
viewing time were evident. In some studies, people were
asked to guess how much time was passing as it happened.
A stopwatch was started, and they had to estimate when a
minute was up. This involves judging time prospectively – as
you go. In the other kind of study time is estimated afterwards – retrospectively. People are occupied with another
task and then asked to guess how much time has passed.
These are two very different skills and I propose that it is
the existence of these two types of time estimation that
causes the Holiday Paradox – the contradictory feeling
that a good holiday whizzes by, yet feels long when you look
back. When life is going smoothly these two types of time
estimation match up, making it feel as though the days and
weeks are passing at normal speed. There are markers in
time that help us to pace the day, such as the start and end
of the working day, the lunchbreak, a favourite TV
programme and bed time. The days are governed by routine
and even variety fits into the pattern in certain places and
involves a fairly predictable number of new experiences (the
six to nine new events we recall from the last fortnight).
Prospective and retrospective time estimation are in synch.
Time feels steady. Everything is in order.

Then you go holiday and these two types of time

estimation fail to tally, causing time to warp. Every sight and sound is new. You don't feel bored. Clock-watching is rare and the familiar markers of time are faint, if present at all. On the same day I was up early to bird-watch in Costa Rica, I can remember more than a dozen experiences – arriving back for breakfast before most people were up; walking into town along the beach and jumping across small water courses; hiring bikes which you had to pedal backwards to work the brakes; searching for a beach we'd heard about and never found; watching a couple having their first surf lesson; cycling along bumpy tracks to a sloth sanctuary where baby sloths were brought out to meet us; a young monkey leaping onto a man's head and leaving a long, greenish-brown deposit on his shoulder; then spaghetti for lunch; a search for tiny, red poison-dart frogs in a botanical garden; followed by a drink in a bar overlooking a surf break known as the cheese-grater because it's so dangerous that the chances are you and your board will be hauled across the coral. I've not even reached mid-afternoon and have come up with more memories than you would expect in a normal fortnight. And this was just one day of the holiday. There were another nine, each with their own new memories. I was so busy that prospective time estimation told me that the day was going past fast. But back home I now use retrospective time estimation when reliving that day, and – because it was packed with new experiences – it seems to have lasted for ages. Employing the memory effects I discussed before, I am using the quantity of new experiences to gauge how much time has passed. I can remember every individual day, unlike in life at home where the experiences merge. Adding up all these new memories makes the holiday feel long overall.

We constantly use both prospective and retrospective estimation to gauge time's passing. Usually they are in equilibrium, but notable experiences disturb that equilibrium, sometimes dramatically This is also the reason we never get used to it, and never will. We will continue to perceive time in two ways and continue to be struck by its strangeness every time we go on holiday.

You can apply prospective and retrospective time estimation to other curious mysteries of time too. Why is it when you're ill the days drag, but when you look back it feels as though the time went fast, almost as if you were never ill at all? Here we see the Holiday Paradox in reverse. Think back to the last time you were unwell – not with something so serious or agonising that you had to go to hospital or feared for your life, but with an everyday illness like a bad cold. The minutes and the hours feel interminable. You long for the day to finish, in the hope that you might feel better next morning. You imagine how fantastic it would be to feel well, and how you will appreciate every moment when you do. You are experiencing time prospectively, wondering when your suffering will end. Your sense of prospective time tells you that every minute is long. All the factors which decelerate time are present. There is no fun. There is no novelty. There is nothing to distract you from paying attention to the clock, the ultimate marker of time. And there is plenty of repetition, mainly of the experience of feeling awful. But once you've recovered, yet again something strange happens – it is the opposite of the Holiday

Paradox, but the cause is the same, the dual perspective on time. Retrospective time estimation kicks in and – looking back – the week you spent in bed feels inconsequential. You can remember feeling ill, but there is such a lack of variation in your memories for that time that the days merge and the period of time barely seems to have featured in your life.

Thomas Mann's descriptions of life at the Swiss sanatorium are a perfect example of the holiday paradox in reverse. He writes that vacuity and monotony 'are capable of contracting and dissipating the larger, the very large time-units to the point of reducing them to nothing at all'. Tedium, he describes as an abnormal shortening of time. He got it exactly right when he said, 'When one day is like all the others, then they are all like one; complete uniformity would make the longest life seem short.'[76]

The other situation where the holiday paradox pertains in reverse is for the parents of small children. The nineteenth-century psychologist and philosopher William James observed that although the years accelerate as we get older, the individual hours and days don't necessarily feel as though they pass any more quickly. Parenthood is a perfect example of this. There is no enforced idleness and definitely no time to sit resting wrapped in blankets, but the result is similar. With early starts, prolonged tiredness and the necessity to repeat tasks and stick to routines, prospectively the days can feel very long. But when you look back on the week, many of the memories are repeats of earlier experiences – you have washed the children, fed them, changed them and read the same book to them hundreds of times before – so the months flash by, underlined this time by the very visible marker of a growing child.

Parenting has the compensation of a new and exciting relationship and the fascination of watching a child grow up. Real boredom is different. One summer holiday as a teenager I worked in a ceramics factory. I had naively assumed my job might be to paint patterns on ceramic bowls. Instead I sat all day at a wooden table which had a metal clamp attached to it and a thin slot in the front of the clamp. My job was to post two-inch-long flat, cream, ceramic oblongs through the slot. Most of them fitted, but a couple of times an hour came the only interesting moment, the moment when I had found an oblong which didn't go through the slot – a dud. It was impossible even to know whether this was a useful job, since no one knew what the rectangles were for. I asked the supervisor and the question went up the line of command until a man came over to ask, in an alarmingly Dickensian way, who was the girl who wanted to know what these pieces were for. Had this been a film, he would have decided this was the sort of enquiring mind he wanted running his company and would then have changed his will in my favour because he had no heir and was looking for a successor to head the family firm. It wasn't a film, and so he didn't. But he did tell me that these ceramic oblongs were to become insulators in washing machines. Sadly this piece of information didn't actually make the job more interesting or time go any faster, which is probably why no one else had bothered to ask. My colleagues had accepted that this job was boring and that all you could do was wish away the hours until it was time to go home. We clocked in and out with punch-cards. You were fined 15 minutes' pay if you arrived more than one minute late and half an hour's pay if you were two minutes late. I soon learnt from the others to make the most of the location of the factory at the bottom of a steep hill. If you cycled down the hill as fast as possible, braked sharply outside the door and threw your bike down, you could clock in just in time and then return to lock your bike up, spending a good 10 minutes chatting while you did so. On the first day all the other women stood up and joined a queue by the door 45 minutes before the shift was due to end. I assumed they were working different hours from me, but they were in fact queuing to clock out. Everyone in the queue watched the second hand moving towards the 12 on the vast clock high up on the wall of the factory. The first person was poised with her card in the air, ready to dunk into the machine with satisfaction on the dot of 6.30 p.m. The company's rigidity over time-keeping had back-fired, leading them to lose almost an hour's work from every member of staff every single day.

In terms of time perception, the Holiday Paradox in reverse was definitely at work here. The hours passed very, very slowly. The marker of time, the big clock, hung over us psychologically as well as physically. We could talk all day and listen to our Walkmans while we worked, but time crawled by at such a snail's pace that we often wondered whether the clock had stopped. Now that I'm lucky enough to work in a job that's never boring, I only ever look at the clock in dread of an approaching deadline and never in the hope of finding more time has passed. Although we wished away the hours in the factory, when we finally reached the weekend and looked back, with so few new memories to fill it, the week occupied little space in the memory and it felt as though it had been short.

I've illustrated how the Holiday Paradox and its opposite can account for some of the contradictions in the sensation of time passing when you are ill, bored, caring for small children or on holiday, but the same principle of our dual perspective on time can also expand on the explanations I have covered so far for that other big mystery we have been considering – why time speeds up as we get older.

Let's take a child of seven who is living a life filled with new experiences. We know that time feels slower for that child than it does for an adult. To understand the reason, once again we need to look at their prospective and retro-spective estimations of time passing. Here there's less of a paradox than there is for adults because even prospectively some hours can drag. Children have far less control over their lives and spend more hours doing things they don't want to. Think back to those endless car journeys or the doodles you did while aching for a dull lesson to end. Conversely, when they're doing something they enjoy chil-dren become very absorbed, with an apparent ability to live more in the moment than adults do. They can entertain themselves in a paddling pool for hours longer than any adult could, constantly innovating and experimenting. For them this time passes quickly, too quickly. They are shocked when they're called away to have their lunch. The watching parent has probably found time dragging, but for the rapt child time has sped by. As bedtime approaches, the minutes rush by even faster as they beg for just one more game/go/story. What the child is experiencing is a variation on the Holiday Paradox, an effect that is complicated by their relatively poor skills at prospective time estimation. The days are full of new experiences and while their parents rush them to school they want to take every opportunity to explore the world. They will stop and stare at workers digging up the road; they will pause to pat a dog; they will notice anything that's different; they will try new things. Why walk along the pavement when you can hopscotch along avoiding the cracks in the paving stones or pick your way up and down the crenellations on a wall? This means that overall, despite a few slow hours where they're forced to do something boring, on the whole days for children, just like ours on holiday, are all-absorbing, and packed with new memories which, looking back retrospectively, makes the months and years seem to stretch out.

By the time a child reaches the mid-teens, the reminiscence effect begins to come into play. The demands of school and exams mean that the hours can still sometimes drag, but gradually there is less routine, more freedom, and novelty in spades – first sex, first drinks, first love, first time away from home, first chance to have some real choice about what they do and who they are. The formation of identity makes these events stand out as we've discussed, giving rise to the reminiscence effect. It's already been suggested that these memories might be extra-strong in order to help reinforce that new identity, but I propose that this time of entering adulthood also becomes the benchmark

for our judgements of retrospective time. This plethora of
new events continues until at least our mid-twenties, by
which time we've become accustomed to a certain number
of memories representing a certain amount of time passing.

In middle age prospective estimation of time tells us that
the *hours* are passing at an average speed and so are the
days. It is the elapsing of the months and years that people
say they find shocking, never the hours. Markers of time
constantly remind us that the years are moving on. We are
shocked when we hear that it is already the twentieth anniversary of the Berlin wall coming down. We see items *we
own* in vintage shops. Most shocking of all, we have work
colleagues born in the 1990s – surely they should still be
at school! These markers in time conflict sharply with our
retrospective judgements, through which we gauge time
passing via the number of new memories we have made.
With fewer new experiences, and thus fewer new memories,
we repeatedly experience a disconnection between the information we are getting from prospective and retrospective
time estimation.

This dual process of prospective and retrospective time
estimation is key to many of time's mysteries. Once again
this is not something to which we become accustomed or
ever will; it is simply a consequence of the dissonance
between the two methods we have for estimating time. We
can't stop judging time in this way, but we can make use
of the features of time estimation to make time feel as
though it's passing slower or faster, depending on how we
would like it to be. This is something I'll be exploring in
the final chapter of this book. Before that, we're going to

move forward into the future. We have seen the way our
memories of the past affect our view of time. Next we will
see how our ability to time-travel mentally into the future
has a bigger impact on the present than we have ever realised.

In case you have resisted looking until now, here are the
correct dates for the list of events:

John Lennon is shot dead – December 1980

Margaret Thatcher becomes
British Prime Minister – May 1979

Chernobyl nuclear
power plant explodes – April 1986

Michael Jackson dies – June 2009

The film *Jurassic Park* is released
in the USA – June 1993

Argentina invades the Falkland Islands – April 1982

Morgan Tsvangirai is sworn in as Prime Minister
of Zimbabwe – February 2009

Hurricane Katrina strikes New Orleans – August 2005

Indira Gandhi is assassinated – October 1984

A car bomb explodes next to
Harrods in London – December 1983

The first cases of swine flu hit
Mexico – March 2009

The Berlin Wall comes down – November 1989

Prince William marries Kate Middleton – April 2011

An IRA bomb explodes at the
Grand Hotel in Brighton – October 1984

Barack Obama is inaugurated
as President of the USA – January 2009

Princess Diana dies – August 1997

Bombs explode on the London
Underground – July 2005

Saddam Hussein is executed – December 2006

33 miners become trapped
in a mine in Chile – August 2010

The first Harry Potter book is published – June 1997

REMEMBERING THE FUTURE

AN ELDERLY MAN, apparently very ordinary, died late one December afternoon in Windsor Locks, Connecticut, in 2008. He was 82. In normal circumstances, the passing of such a man would have provoked little interest beyond his family and friends. But in this instance, a team of internationally eminent research scientists from across the USA immediately sprang into action. The care home where the man lived – and died – phoned Suzanne Corkin, a neuroscientist at MIT in Boston, who – for once – was not away at a conference. She alerted one of her colleagues, based in California, who was also in the country – and he too picked up the call. So keen were the scientists to get hold of the man's body as soon as he died that they had even taken the precaution, years in advance, of contacting every funeral home in the area. If his mortal remains were delivered to them, they must on no account proceed with cremation.

Why were they so insistent? In order to stop the scientific calamity of the most famous brain in neuroscience being incinerated.

This outcome averted, ice blankets were packed around the elderly man's head and his body was driven in a hearse to Boston a hundred miles away. Meanwhile Jacopo Annese, a neuroanatomist was on a plane from California. He had been selected by Suzanne as the perfect person for the job – someone advanced enough in their career to be highly skilled, yet prepared to devote the necessary time to the project. By midnight scans of the brain while it remained inside the skull had been completed. Next morning Suzanne watched through the viewing window of the pathology department while Jacopo and two colleagues carefully extracted the brain from the body. The following day Jacopo flew back to San Diego and 'Henry', as he couldn't help saying, occupied the seat next to him.

It might seem rather tasteless to give a brain, hanging upside down in formaldehyde in a plastic vat inside a cooler, a name. But for the scientists it was somehow appropriate. Professor Corkin had taken responsibility for Henry Molaison, a vulnerable soul in life, and ensured he was well looked after at the Bickford Nursing Home. She had protected his identity – for decades in text books he was known only as 'H.M.' – and grown very fond of him. But in scientific terms, it was his brain that made Henry so special. For more than 45 years, Suzanne had thought more about the workings of this brain than any other – what it could remember, what it could learn, what it could predict. Now finally she and Jacopo were to have the opportunity to see inside it.

Why were they so excited? Because Henry had lived two-thirds of his life locked in an eternal present.

When he was 27 years old, Henry underwent brain surgery to try to prevent the multiple epileptic seizures he experienced daily – seizures that would probably have killed him within a few years if no action had been taken. The surgeon, Dr William Scoville, took a silver straw, inserted it into Henry's brain and slowly sucked out part of the hippocampus, the tiny seahorse-shaped area deep inside. The operation appeared to be a success. Henry recovered well and the seizures ceased. But Scoville gradually realised that something had gone very wrong. Henry could not remember anything new that happened to him. Although he could recall incidents from his childhood, people he had met the previous day were like strangers to him. Every face was new. Every experience was new. He had no idea what he had been doing even an hour before. The surgery had caused anterograde amnesia. Henry had his old memories, but he could never again make any new ones.

Although Henry's case features in many neuroscience and psychology textbooks, it usually appears in the chapter on memory. There is, though, another aspect to amnesia, one that's less well-known: the loss of the ability to imagine the future. Henry exhibited just these twin symptoms: no sense of a past after the accident, but no sense of a future either.

As I've illustrated throughout this book, we construct our own sense of time in our minds, but nowhere is this clearer than with our mental images of the future. At will, we can choose to imagine tomorrow, next week or 1,000 years hence. This ability has nothing to do clairvoyance. The chances are that the future we imagine will not turn out exactly that way. Indeed, such is the power of our imagination that we can clearly picture a future that not only won't happen, but could *never* happen. This strange ability to pitch ourselves forward in time – called future thinking – is the opposite of memory. But as I'll show in this chapter, it is connected to it. For the mind uses both our sense of space and our memories to create a sense of the future.

On average people think about the future 59 times a day or once every 16 minutes during waking hours.[77] Indeed research in this field reveals a staggering finding: contemplating the future could be the brain's *default* mode of operation. But this is not wasteful daydreaming or, as the phrase goes, a case of 'wishing our lives away'.

Mental time-travel into the future matters – and is useful. It affects our judgements, our emotional states and the decisions we make, sometimes for the worse. And my explorations of future thinking will also reveal something rather surprising about the reasons for the fallibility of our memories of the past.

TIME-TRAVELLING INTO THE FUTURE

There has been more than a century of research into the way our memories work, but future thinking is a far newer area of study. The most significant finding is the degree to which future thinking relies on mental time-travel in the opposite direction – into the past. This can even explain one of the mysteries of memory – why it so often lets us down, why researchers like Marigold Linton found studying her own memory to be so painful. We need our memory to be a reconstructive process, and one that's flexible and even unreliable, to allow us to imagine the future. The evidence for this idea comes from a range of sources, beginning with patients like Henry. Hundreds of cases of amnesia have been reported in the medical literature and their doctors have often noted that these patients found it difficult not only to remember the past, but to imagine the future. They are unable to picture what they might do the following day, let alone in a decade's time. Although many doctors have made this observation, future thinking has only been studied systematically in a handful of these patients, compared with the copious amount of research done on memory skills.

There once was a man known in the medical literature as 'N.N.', which stood for 'No Name'. He was driving up the exit slip-road from a motorway in 1981 when he came off his motorbike and sustained a serious head injury. Like Henry, N.N. doesn't remember new information and still reacts with horror every time he hears what happened on 9/11.[78] He later became known by his real initials K.C. and was visited by the influential memory theorist Endel Tulving. Tulving was famous for differentiating between semantic memory – our memory for knowledge, so for example that Canberra is the capital of Australia; and episodic memory – the memory for the personal events that happen to us, for example visiting Canberra. Tulving asked K.C. some simple questions: 'What are you going to do tomorrow?' and 'What are your plans for the summer?' K.C. was unable to answer either. When Tulving asked him what was in his mind, he replied that it was blank. Hard as he tried, he could not hold ideas of the future in mind, and 30 years later he still can't. Some patients, such as one known as D.B., can imagine political events in the future, but find *personal* future events particularly difficult to envisage.[79] It's striking that not only do these patients struggle to imagine a future when asked to do so, but they appear to have no desire to imagine one either.

Just as semantic memory and episodic memory are different, a similar distinction applies to the way we think of the future. There is a difference between sitting in midwinter *knowing* the fact that it will be warmer when summer comes and *imagining* yourself sitting in the sunshine next summer feeling the heat on your skin. This latter mental time-travel into the future through imagination has been defined as episodic future thinking, which for simplicity I refer to here as future thinking. It is part of a more general system of mental time-travel that Endel Tulving said makes up our autonoetic consciousness. This is the sense we have of ourselves as persisting across time, and it is manifest in our ability to re-experience or pre-experience events. This pre-experiencing involves imagining what an activity might feel like, not merely intending to do it. So you might notice in your diary that you're meeting some friends for a pub lunch. It is your prospective memory that reminds you to turn up at the right time, but it is future thinking that means you can picture yourself ordering a drink, finding a table and reading the blackboard to see what specials they have. With future thinking you project yourself forwards mentally to imagine the actual experience. This is different from actively planning, and it is this skill that seems to set us apart from other animals.

Mental time-travel need not involve vast stretches of time. Often it will concern what you have just done or are about to do. If you take the example of a job interview, there's the mental rehearsal of the questions beforehand, followed afterwards by the agonising, repeated replaying of the worst moments, coupled with the imagination of a possible past, of the things you wished you'd said. These were possible futures. Now they are part of an impossible past.

Demis Hassabis and Eleanor Maguire were the first neuroscientists to carry out a structured investigation of future thinking in people with brain injuries. They found that even when they gave them suggestions for all the sensory details that might be present in a future scenario – the sights, the smells and the sounds – they were still unable to imagine the scene.[80] These five patients had a range of scores on tests of IQ and memory skills, but when it came to imagining the future four of them did very poorly, despite the fact that the scenarios used were everyday enough not to appear to necessitate access to detailed memories.

It is not only people with brain injuries who find future thinking difficult. Anyone with a poor autobiographical memory will also find it harder to project themselves into the future. This includes very young children, people with schizophrenia, Alzheimer's disease or depression and those who are feeling suicidal.[81] The more delusions and hallucinations a person with psychosis experiences, the harder it is for them to hold an idea of the future in mind, depriving them of the agency that mental time-travel can bring.[82] As the decades pass we discuss our failing memories, but this is accompanied by something we tend to notice less – a decline in our ability to imagine the future. This lends weight once again to the idea that we rely on past memories in order to conjure up ideas of time yet to come.

The part of Henry's brain that was most damaged was the hippocampus. It gets its name from its shape – it curls

around in a narrow arc similar to a seahorse. Not long ago
I stood on an old converted squash court that now houses
a brain bank. There were shelves and shelves of brains in
vats. The neuroscientist carefully handed me one of their
six thousand brains and pointed out the hippocampus. It
was extraordinary to think that an area that is only four
centimetres long and formed so delicately could hold the
key to the lifetime of memories that form a person's identity. We know this area to be crucial to memory, but the
experiences of the patients with amnesia would suggest that
it also plays a part in imagining the future. The evidence
from these people begins to build up a picture of the way
the brain holds ideas of the future in mind, one which is
backed up by scans of living brains.

To understand this better, think of something you know
you are going to do next week, but not something that
happens every week. Now try to come up with a detailed
picture of this event. If it's indoors, what will the room look
like? If there are people, what will they be wearing? Look
closely at the details you have conjured up, apparently from
nowhere, and you will probably find they involve memories
from your past. I know that I'm going to Oxford next week
to interview a psychology professor about his work on
group cohesion. I've never met him in person, nor been to
his office, yet I can summon up a rough picture of a scene
where we sit on faded velvet armchairs in his wood-panelled
office. The desk is covered in piles of paper and as well as
floor-to-ceiling bookshelves along one wall, there are stacks
of books on the carpet too. In fact this picture seems to be
a combination of what I saw the last time I went to meet

an Oxford professor and scenes from films like *Educating Rita*. I could be wrong of course and his office might be modern and minimalist. Nevertheless it is clear that my vision of the future is an amalgamation of any relevant memories I can find. By recombining old memories we are able to project ourselves forward into the future, giving us endless combinations from which to select the most plausible possibilities. Like a remix, utilising these memories allows us to preview future events in a window in the mind.

Considering this, it is not surprising that if you scan a person's brain while asking them to imagine a future event, one of the areas that shows activity is the hippocampus, the repository of memories, the very same area that was sucked out of Henry's brain with a straw. In fact the regions of the brain used to recall the past largely overlap with those used to imagine the future.[83] Memory is essentially a reconstructive process; when we want to re-experience an event we don't summon up a tape from the library. We reconstruct it and even alter that memory if new information has changed our views since it first happened. A similar process takes place when we imagine the future.

The neural signatures of remembering the past and imagining the future are remarkably alike. Researchers have investigated this by giving people a key word or phrase and asking them to imagine different past and future scenarios involving that word. One researcher, Karl Szpunar, gave people the name Bill Clinton.[84] People reported finding the task easy (maybe it's just me, but one particular Oval Office scene does come to mind, so I hope this didn't skew their study). Three main regions of the brain were involved both in thinking about the past and the future. The first is the frontal lobe. This is the area behind our foreheads that houses working memory and is responsible for making decisions and solving problems. The frontal lobes also ensure that memories from the past are not mistaken for real life.

The second area is the parietal lobe. These are found one on each side towards the back of the head, at the top. It is here that sensory signals from the rest of the body are processed and it is also where letters are combined into words and words into thoughts. Intriguingly, part of this area also allows us to find our way around the world. This navigational function sheds more light on the shared mechanisms for thinking about time and space that I was discussing in Chapter Three. It suggests that we might conjure up the past by imagining pictures in the mind's eye, just as we do when we're trying to find our way to a place we have visited previously. The patients with amnesia found it particularly difficult to imagine the spaces in which new events might take place.[85] When asked to imagine a future scenario where they were standing in the main hall of a museum, people without brain injuries tend to describe features such as a marble floor, a domed ceiling or paintings on the wall, but the people with amnesia were unable to add these elements to their vision of the room. They didn't find the task any harder and even considered their imaginary room to be quite realistic, but were less likely to mention objects, feelings, senses or anything to do with the location of themselves in the room. Somehow the spatial context was missing, backing up the idea that we use space in order to construct a sense of time in our minds.

The third area activated by thinking about either the past or the future is the medial temporal lobe. This contains the all-important hippocampus, and regulates memory, learning, language and emotions. Although the same parts of the brain are used, imagining the future takes more brain power than remembering the past. I love the fact that the further into the future we transport ourselves, the more active the hippocampus becomes. It is intriguing too that some of the brain regions used to imagine the future are also employed when we consider what someone else might be thinking, when we try to simulate their state of mind. This suggests that during future thinking we are doing something similar – simulating our own state of mind in another time and place. There is still much to be learnt about the workings of this process, but it does appear to be something that's uniquely human.

CAN YOUR DOG PICTURE NEXT WEEK?

If you have ever owned a dog it is tempting to believe that in idle moments your dog thinks back fondly on the best walks you have ever had together, reminiscing about the time they found a dead rabbit, the afternoon where they were allowed to run wild in fields chasing other dogs, the occasion where they managed to strain enough at the lead to reach some chocolate buttons caught under a shop counter. Unfortunately it is unlikely that dogs recall these happy times at all. They can remember how to navigate their way to a favourite patch of grass, and will tug at the lead to get there, but as far as we know they have no way of remembering the individual events that occurred there. This would suggest that they can't time-travel into the future either. They can't decide to picture themselves on Christmas Day lying by the fire gnawing a bone. So if dogs can't do it, what about other animals that are known for their intelligence?

Panzee is a female chimpanzee who is clever enough to distinguish a pint from a half pint. Using a keyboard she can identify certain foods and objects and after years of training from Charles Menzel, an anthropologist at Georgia State University in Atlanta, she has learned to use 256 different symbols. Yet she displays no evidence of future thinking. She can point to the place where she has hidden some food previously, but this demonstrates that she has memorised the location, not that she can remember the action of hiding the food or picture retrieving it. There are of course animals that appear to plan ahead. How about the impressive accuracy with which squirrels return to the precise spot to dig up the nuts they buried there several months earlier? You could interpret this as evidence both of a reliable memory and an understanding of their future needs during the changing seasons. However, some naturalists now believe that squirrels simply start digging in typical hiding places and have no idea whether they are finding their own cache or one buried by another squirrel. You could argue that by hiding food at all they are planning for the future, but an instinct to hoard food is not the same as *imagining* themselves feeling hungry in the future and making plans accordingly.

Of the animals that have been studied, it is a bird called the western scrub jay which comes closest to having the human concept of a past and a future. These birds not only *look* smart with their shiny, blue plumage, they *are* smart. Native to North America, they belong to the same family as rooks, crows and ravens and are among the most intelligent of all birds, but it is their tendency to hoard that interests comparative psychologists like Nicola Clayton at Cambridge University. If they experience a food shortage they learn to gather food and hide it for later. In experiments conducted at Cambridge University, Nicola Clayton has found scrub jays will even plan for the future when they have everything they need in the present.

Nuts stay fresher for longer underground than dead worms. Scrub jays know this and will select caches accordingly, demonstrating that they not only remember where they hid the food, but what it was and how much time has elapsed since they created the cache. They even remember which other birds were watching them when they hid the food and will re-cache if they know they were observed by other birds and have previously stolen food themselves (not all of them do). This suggests they are utilising their experience rather than relying on instinct alone to plan for the future and it provides compelling evidence for their planning skills and reliance on memory. The latest research from Costa Rica has revealed that 21 other species of bird, from the orange-billed nightingale thrush to the white-whiskered puffbird, might have similar skills. It has just been discovered that in a practice known as bivouac-checking, these birds learn to check ant nests (or bivouacs) at the end of a day in order to follow the army of ants the following day when they go out on raids, sweeping through the forest to drive out insects.[86] This appears to be evidence of memory and planning.

But again the question has to be asked as to whether this demonstrates future thinking in the human sense. In order to hoard or re-cache food the birds do not necessarily have to *imagine* themselves in a future where there is none. Once again it is important to distinguish between knowledge of the past and future and actually re-experiencing the past or pre-experiencing the future. If I ask you to imagine where the scissors are located in your house, there is a difference between you picturing them lying in a drawer and remembering the action of putting them back there the last time you used them.

Another very intelligent animal, the dolphin, shows signs of the ability to time-travel mentally into the recent past. Dolphins can be trained to 'do something they've not recently done', so at a given signal they will perform a trick they have not done for some time. This suggests that they do have some recent autobiographical memory, but again the evidence of true mental time-travel in animals just isn't there. Some people find this disappointing; we seem to like the idea that animals have memories and imaginations like ours, especially when it comes to our pets. A psychologist who has done a great deal of work in this field, Thomas Suddendorf, even apologises for being such a killjoy.[87] And it's not just animals who are so deprived. Babies are forced to live in the here and now too, unable to escape mentally into the future. It is not until the age of three or four that they begin to be able to imagine a future where they might feel differently, where they can antic-ipate or fear events. This imagining helps them to begin to develop the crucial skill of emotional regulation, where they find ways to control their emotions. Adults can comfort themselves with the knowledge that the agonising pain from a badly stubbed toe will not last forever, because they know this from experience and can easily imagine a future where their toe doesn't hurt at all. Babies are stuck in the present, unaware that they might feel differently in the future.

WHAT ARE YOU DOING TOMORROW?

Before I begin an interview for a radio programme, I need to ask the person a question to get them talking enough to check the sound levels. The classic question is, 'What did you have for breakfast?' but too many people give one-word answers like 'nothing' or 'toast', responses too short to be of any technical use. So I like to ask people what they are planning to do that afternoon or the following day. Last week a woman told me she was going straight back home because two tree surgeons had arrived at her house with chainsaws to sort out a tree in her garden, but appeared to be drunk. She was keen to get back to discover what would be left of her garden, or even of the men themselves. The answers I get are rarely this dramatic. It is an easy question, or it is if you are an adult. If you are three years old it is much more difficult. In one experiment only a third of three-year-olds could give a plausible answer as to what they might do the next day, but within a year or two their sense of the future has developed to the extent that two-thirds of them can do it.[88]

With any tests on young children there is always the issue of whether it is really their thinking that is holding them back or their verbal ability. Can we be sure that they understand the wording of the question? Most three-year-olds know that tomorrow is in the future, although they don't always realise that it denotes the next day. What they do know is whether or not they like pretzels. So to overcome any verbal difficulties the psychologist Cristina Atance gave a group of children some pretzels to eat. Once the salt had made them thirsty she gave them a choice of more pretzels or a glass of water. Most kids chose water. But when she asked them what they would like to have the next day, while most adults opt for more pretzels, most small children still chose the water, unable to pitch themselves mentally forward into a future where they would no longer be thirsty and would be keen for more pretzels.[89] If you think back to the areas of the brain involved in imagining the future, it is not surprising that toddlers find it difficult. Two out of three of these areas, the parietal and frontal lobes, do not develop sufficiently until the second or third year of life. This suggests that small children have an extreme form of the empathy gap that we can all experience on occasion, which leaves us unable to imagine that we might feel differently in the future. If you're packing to go on holiday to a hot country, but at home it's snowing, it is tempting to pack plenty of socks and a jumper because it's so hard to imagine ever feeling too hot.

MEMORIES FOR EVENTS THAT NEVER HAPPENED

Ten years ago I took part in an experiment on pain tolerance that involved submerging my arm into a bucket of iced water and holding it there for as long I could. At first it seemed fine and I thought I could handle the discomfort. It was only cold water after all. Then a deep ache gradually spread up my arm and the pain got worse and worse. I can still remember the feeling of all-encompassing agony that was impossible to ignore. I could stand just 90 seconds before I pulled my freezing, numb arm out of the water. There is just one problem with this memory; it didn't actually happen. Until recently I was convinced that it did. I had recorded myself taking the pain test, along with a male and female volunteer, for a radio programme on sex differences in pain tolerance. When I was making a new programme on pain relief this seemed like the perfect piece of archive to include. I found the tape and handed it to the producer, describing to her the torture of holding my arm in that bucket and how I'd discovered that I had a low pain threshold. The patient producer listened to an hour of audio, searching for the part where I put my arm into the icy water, but it turns out that I didn't do it. I had bravely recorded two other people taking the test, without doing it myself. I am convinced I can remember exactly what that pain was like, but I recorded the whole session and so the evidence is there, or rather isn't there, on the tape. It didn't happen.

This fallibility of memory is uncomfortable, but it could be the consequence of our ability to imagine the future. The fact that memories are so crucial to future thinking could explain one of the long-standing mysteries of memory – why it so often lets us down. Elizabeth Loftus has become one of the most eminent living psychologists for her demonstration that memory is not like a videotape. These are some of my all-time favourite experiments. They're relatively simple, yet so cleverly constructed that they have had a huge impact on the way the law courts consider evidence from eye witnesses in criminal trials. Loftus succeeded in implanting false memories in people's minds, not through hypnosis, but simply by convincing them that they remembered events that had never taken place. Using interviews with relatives to give her some background information about true events, she would then discuss the past with an individual and make them believe they had been lost in a shopping centre as a child or remembered kissing a giant green frog or that they met Bugs Bunny at Disneyland. This might sound plausible until you recall that Bugs Bunny is a Warner Brothers character, so he's never going to be allowed into Disneyland. Our memories are flexible. They don't enter our minds perfectly formed and then remain intact in the mind's archive, waiting for us to call them up. Decades of evidence demonstrates that we change memories as we lay them down, we alter them again if new information comes to light, and then, if it helps us to make sense of events, we change them again when we recall them. Memory is reconstructive. None of this is done dishonestly or even consciously. But while the flexible nature of our memories presents a problem for the reliability of eyewitness testimony, this same flexibility could be the key to us imagining the future at all.

If memories were fixed like videotapes, then picturing a new situation would be time-consuming. Let's pretend that you want to picture yourself arriving by double-decker bus at a tropical beach for the wedding of Johnny Depp to your best friend. In an instant you can do this. If memories were rigid this would be a complex process. You would need to do the equivalent of finding your personally taped memories of sitting on buses and visiting your best friend, and then you'd need to order up clips from the mind's archive of films starring Johnny Depp and TV programmes featuring tropical beach weddings. These memories could be years, even decades apart. Once you have extracted all the necessary elements you'd then need to splice them together to invent this scene. Cognitively it sounds like hard work and would be if we had to do it this way. In fact the flexibility of our memories makes it relatively easy because we can meld all these different memories together seamlessly to invent a new imaginary scene, one which we have never even contemplated before, let alone witnessed. The flexibility of memory seems to be the key to imagining a future.[90] Our millions of fragments of memories from different times of our lives are not set in stone; they can change, giving us endless, instant imaginative possibilities. Our unreliable memories might feel like a deficit, but they facilitate mental time-travel into the future.

It is obvious that we learn from experience, but taking this one step further perhaps the primary purpose of memory has nothing to do with looking back, but more to do with allowing us to look forward and imagine possible futures. This is not a new idea. Medieval illustrations of the mind from the fourteenth century depict memories like snakes feeding into the imagination and, long before this, both Aristotle and Galen described memories not as archives of our lives, but as tools for the imagination. It was in 1985 that the Swedish neuroscientist David Ingvar proposed the modern version of this idea. Since then there has been a flurry of studies on future thinking, although, as I've mentioned, this research is still dwarfed by the attention given to memory.

In some ways imagination is easier to study than memory

because you avoid the problems I was discussing in the last chapter, such as the necessity to check the accuracy of the memories. The beauty of the imagination is that you can ask every person in your study to picture the same thing.

> Take the word 'forest' and think of a past memory you'd associate with it. Take a moment to imagine what you can see, what you can smell, whether you feel cold, whether you are happy or sad, who you are with and what you are doing. Now imagine a future event in a forest. What's that forest like? Is it dark? Does it smell nice? Are you with anyone else? What emotions are you experiencing? Now compare these two images. Which is the most vivid?

Laboratory studies have found that memories for past events tend to be more graphic than thoughts of the future, and they include a greater number of sensory descriptions of how things looked, sounded or smelt. Yet we know that despite the lack of detail, cognitively it is still more demanding to imagine the future than to recall the past. It is often argued that the distant future might be sketchier because imagining it in detail would be a waste of cognitive resources. I wonder whether it is simply that we don't have the information. I can easily imagine having lunch at home in a month's time, but to imagine lunch at home in a decade's time is much harder, since I don't know where I'll be living or what my surroundings might look like. Researchers have also found, rather neatly, that just as memories of last week are more vivid than a decade ago, so a corresponding effect happens with future thinking; events in the near future can be imagined more acutely than those in the distant future.[91]

Although the past gives rise to more vibrant descriptions, when it comes to our emotional response, it is the future which is the more potent. Research shows that anticipation evokes stronger emotions than retrospection, whether positive or negative. For some the anticipation of a holiday can be almost as good as or even better than the holiday itself. Future imaginings are on the whole more optimistic and also more personal.[92] Most of us believe than just a month from now we might have more money, and the further into the future we look, the more optimistic we become. Gamblers believe they'll be luckier in the distant future, placing safe bets now and long-shots for further ahead. When students were asked to list 10 important events from the past and 10 from the future, the future events were more positive.[93] You might guess that this optimism is a reflection of their age, but even up to the age of 75 most people believe their future will be better than their past.[94] And if you ask people to think deliberately of negative events that might happen in the future it takes them longer to come up with those than negative events from the past.

The question remains, why emotions should be stronger when imagining the future than recalling the past? The answer could revolve around the idea of uncertainty. We know that this gives rise to more intense emotions and inevitably the future is less certain than the past. But what if the future event doesn't involve any uncertainty and is definitely nice?

Imagine you have just opened an envelope. Inside is a letter saying you have won an all-expenses-paid skiing trip for two to the Whistler-Blackcomb Ski resort in Canada. The prize includes flights, a lift pass, ski hire, a skiing lesson and five nights in a suite with a jacuzzi at the five-star Chateau Whistler, which is just five minutes' walk from the ski lifts.

Your first thought might be that it's a scam, but the volunteers in this study were told to imagine it was all above board and that they had entered a local radio competition with precisely this first prize. Half the volunteers were given the scenario and asked to rate their happiness as they looked forward to the holiday. The other half were asked to imagine they had already taken the prize holiday. Then they were asked how happy it made them feel. The people who imagined it in the future felt happier than those who were reminiscing.[95] This task involved no uncertainty, suggesting there might be an alternative explanation for the emotional nature of future thinking. One theory of emotions is that their purpose is to prepare us for action, ready to avoid the negative and embrace the positive. This would fit in here. We don't need to be readied for action when it comes to the past, so memories need not have such strong emotional content as thoughts about the future.

Future thinking is clearly useful in terms of planning; it allows us to experiment with hypothetical situations before we make a decision and is key to the extraordinary ability of humans to adapt to their environments. However, the mental rehearsal of a future event can occasionally be so realistic that we become convinced we have actually done it. Just as Elizabeth Loftus implanted false memories in the minds of her volunteers, we do the same to ourselves – like the email you assume you sent, only to discover later that you only *thought* about writing it. Considering that the same parts of the brain are recruited and similar processes are used to conjure up the past and the future, it is perhaps surprising that we don't become confused more often. Some researchers believe that it's the vibrancy of memories that allows us to distinguish them from future thoughts. When we do make mistakes they occur in one direction. We might believe we have already done something we only thought about, but it is very unusual to do it the other way round, to believe that a past memory was merely a daydream.

Alan Johnston's experience as a hostage illustrates just what a powerful impact mental imagery can have. The rest of us might not harness it so consciously, but anyone who has imagined answering questions before a job interview or practised a difficult conversation with their boss is employing the strategy of mental rehearsal. In sport it is now common-place for athletes to be taught skills in visual imagery where they picture every detail of winning. Tennis players learn between-points routines that aim to return them to the perfect frame of mind before each serve, regardless of what happened during the previous point. After a bad line-call a player with the mental strength to win will re-group, forget the past, and project themselves into the future, imagining the ace they're about to serve. Snooker players use imagery to picture the ball gliding straight and fast to the pocket. The only time I was any good at pool was when I was

playing with a sports psychologist who was working with
Olympic athletes. Whenever it was my turn he talked me
through imagining the perfect pot. Remarkably it worked.
After that I looked forward to the next opportunity to dazzle
people with new skills at pool, but sadly, without the presence of the psychologist, my ability to imagine the perfect
pot – and so to realise it – deserted me. This is why top
sportspeople invest time in practising their imagery as well
as their athletic techniques.

Even if you are not trying to win a game, imagery about
the future can help you on an everyday basis. When people
develop elaborate mental images about a future event, they
are much more likely to remember to do it. So if you want
to make sure you don't forget to buy eggs on the way
home you need to picture yourself going into the shop,
standing in the correct aisle, looking for the medium-sized
eggs, opening the lid to check they're not broken and
taking them up to the counter to pay for them. This is far
more effective than saying to yourself, 'Mustn't forget the
eggs.' You can apply this to other situations too. If you
have an exam coming up you can use your mind to improve
your marks, but it is important to do it the right way. In
a study one group of students spent five minutes a day
in the week running up to their exam imagining the
moment when they would discover they'd achieved an A.
Meanwhile a second group spent their daily five minutes
imagining the process of studying for the exam – imagining
finding a quiet place to study and getting things ready.
When the real exam results were revealed this was the
group that did better.[96]

SUICIDE ISLAND

The conscious choice to pre-experience the future through imagery can bring many benefits, but sometimes our minds propel us into the future against our will and the consequences can be fatal. Just 10 kilometres south-west of Hong Kong Island is a much smaller island called Cheung Chau. On the Saturday when I visited the island, the fast ferry was packed with families carrying picnics and beach towels. A sign on the inside of the ferry cabin said, 'Be considerate. Keep your voices down!' Most people seemed too excited to take much notice.

Later on, while I sat on the waterfront promenade eating dim sum, I was struck by the combination of a Chinese scene – stalls selling traditional herbal medicines and tiny fish, laid out in lines on racks to dry in the sunshine; and a traditional British seaside resort atmosphere – children carrying buckets and spades and begging their parents for ice creams. Teenage boys cycled past on tricycles with their girlfriends on the back seats, shaded by blue-and-white striped canopies. The harbour was crammed with so many junks and fishing boats painted cobalt and green that you could barely see the blue water in between them. A short walk away at the recreation ground next to the Buddhist temple men were dismantling the bamboo-covered steel scaffolding after the annual bun festival, which features the famous 'bun climb'. Men scramble up to the top of 20-foot-high scaffolds which are covered in sweet buns to form a giant, sticky pyramid. The climb is so precarious that after an accident in the 1970s, in which one of the towers collapsed injuring 30 people, competitors are

now required to complete a training course in basic moun-
taineering skills. And sadly even the buns are replicas now.

High above the tower of buns, the temple and the harbour,
there are blocks of holiday flats built into the hills, half-hidden
behind the trees. With sandy beaches and alleyways lined with
pastel-painted cottages, this could be the perfect holiday destination. But at the start of the twenty-first century the island
developed a new reputation as the place from which not
everyone returns alive. A small number of people began visiting
the island with a very specific plan in mind, a plan to die. The
tiny community sometimes had to deal with a dozen deaths
from suicide a year. For the locals, it was not only distressing
to find the bodies, but bad for business. The suicide rate in
Hong Kong is four times that of the UK. Experts blame the
pressures of life in a competitive, crowded city combined with
a common feeling that asking for help is shameful.

At Hong Kong's oldest psychiatric hospital, a place now
modernised but so infamous that people still warn each
other jokingly that if they're not careful they'll be 'sent to
Castle Peak', I met Angela, who had come from rural main-
land China with her husband in search of work. She told
me that a life of poverty and discrimination had ground her
down. After her youngest daughter was born she was diagnosed with depression and the baby had to go into foster
care for a year. Angela became convinced she was a bad
mother and told me how in her despair she decided her
children would be better off in another life – the other world,
as she put it – rather than in this one. She made a decision
to kill her children and then herself. Thankfully she told
staff at the hospital of her plans and was given help before

she was able to try. Now in her early fifties; she told me she feels a bit happier and that she and her husband quarrel less, but that still she doesn't have much hope for the future.

Angela's plan was not carried out, but others continued travelling to Cheung Chau to kill themselves until the community took advice from the eminent Centre for Suicide Research at Hong Kong University. Researchers there have found that at least a third of people attempting suicide do so on impulse and have had no previous signs of mental illness. This means that if you remove their first choice of method they might never try it again. As many as 7,000 lives were saved in Britain after the switch beginning in 1958 from coal gas, which kills you if you breathe in enough of it, to the safer natural gas,[97] while the suicide rate in Samoa dropped after the move from Paraquat to other less toxic types of pesticide. The residents of Cheung Chau have tried something similar. As the holidaymakers crowded down the ramp from the ferry, I noticed a couple of police officers standing quietly at the quayside. They were looking out for anyone on their own who appeared vulnerable. If they spot someone they say hello and offer them help. The holiday flats can no longer be rented by single individuals and if owners do become concerned about their tenants they knock on the door, sometimes repeatedly, asking if they can be of service. Police tour the island looking for anything suspicious. The experts say that some of those determined to end their lives will go ahead, but for the community here these steps have made a difference. Suicide is no longer such a problem there.

It is hard to imagine what goes through the mind of

someone like Angela or the suicidal visitors to the island.
Earlier in this book I was discussing how the suicidal state
can skew the perception of time to the extent that it can be
hard for people to envisage a future at all. But very recently
it has emerged that there is one type of future imagery that
suicidal people often do experience – involuntary flash-
forwards to the future. Like a flash*back* after a trauma, the
pictures appear in your mind when you least want them to
and are difficult to banish. Working in the psychiatry department at Oxford University, Emily Holmes has found that
at the time of greatest despair, people who are feeling suicidal
often find that images of their imagined suicide intrude into
their minds.[98] One man repeatedly visualised himself
deciding whether to leave a final message and then jumping
from a specific cliff. The image was so detailed that he could
clearly observe his own feet, the grass and the rocks below
him. Several times he had left hospital and tried to reach
this exact cliff top. A woman was plagued by images of
herself feeling cold and damp inside a coffin.[99] A man
pictured the particular place on his daily journey where he
planned to crash his car. A few found these images
comforting, despite their graphic nature, but for others they
were distressing. This is what one woman imagined when
she was planning to jump off the top of a five-storey house.
'I hit the road and the concrete. I imagine my brain splitting
open like a pumpkin, seeing myself doing that, seeing myself
flying down, hair and clothes flying backwards, head breaking
into pieces, making a sound like a watermelon, a pop sound.
The traffic stops and people scream, my mother comes out
screaming, mother is crying, father is in shock, face so

shattered that it is unrecognisable.' This graphic image haunted her. Some said these imaginings took over their lives.

Research on future thinking suggests that these images could have serious consequences. There is plenty of evidence that once you have imagined yourself doing something in the future, whether voting or giving blood, you are more likely to do it. It is routine for mental health professionals to ask distressed people whether they have had suicidal thoughts, but it is rare for them to ask whether they have experienced unwanted future images like these. The new discovery of these flash-forwards could even be used therapeutically, discussing the images with the suicidal person, but replacing them with different endings where they don't kill themselves, showing that other futures are possible.

This rather grim finding demonstrates the power of future thinking. This situation is extreme, but we all find our minds drifting into the future many, many times a day. This leads us on to the question of whether thinking about the future could even be the mind's default position.

THINKING ABOUT NOTHING

Like many students I had a go at learning to meditate. And like everyone else on my course I bought the postcard that seemed to sum up our difficulties. It featured a black-and-white cartoon drawing of a man sitting cross-legged trying to meditate. The space around him is packed with bubbles of unwanted thoughts. 'I don't think I'm very good at this.' 'Am I thinking about the right thing?' 'My knees hurt.' And then as his mind starts to wander, out come endless thoughts about the future. 'How much longer till I can go?' 'How shall I get home?' 'What shall I eat tonight?' 'What shall I do for Christmas?'

The postcard is intended as an illustration of the difficulties of mastering meditation, but it is also a nice example of something else: the brain's default network. The idea that we only use 10 per cent of our brains is a complete myth. Even when you are lying completely still apparently thinking about nothing, many parts of the brain remain active. This is where one of the most fascinating findings about our conception of the future comes in. All three of the chief areas of the brain involved in imagining the future are part of the default network. It is almost as though our brain is programmed to contemplate the future whenever it finds itself unoccupied. In meditation you are told to sit and monitor your thoughts as they come and go. If you try this, even for a moment, it is hard to escape from thoughts about the future.

Daydreaming might seem like a waste of time. We all fight our own lack of concentration, but with the exception of the rare individuals who are compulsive daydreamers, mind-wandering is a useful skill. There is a good reason why our brains invest so much effort in it. Using thought-sampling procedures, you can calculate the frequency of daydreaming. Researchers at Harvard University used an iPhone application to monitor the wandering minds of 5,000 people living in 83 different countries. The phone pages them at random intervals to ask how happy they feel at the moment, what they are doing and whether they are thinking about something other than what they're currently doing.

It was revealed that a third of the time people's minds were wandering. Sex was the one exception to this, where people claimed to be keeping their minds on the job (yet somehow responding to their iPhone?). Sadly, in contrast to the optimistic nature of deliberate future imaginings, unintentional mind-wandering didn't necessarily make people happy. Half involved pleasant topics but that still didn't make them feel good, and the daydreams that were neutral or unpleasant made them feel unhappy. So imagining the future might have its uses, but when it is unintentional, as the authors put it, 'it comes at an emotional cost'.[100]

So our minds are constantly on the go, imagining possible futures, but why doesn't the brain instead take the opportunity to rest when it has the chance? If we were always focused on upcoming events and concrete plans, this tendency to dwell on the future might make sense, but often we conjure up situations that would be life-changing but highly unlikely – so why do we do it? Daydreams can undoubtedly help us to plan for future eventualities, but Moshe Bar from Harvard Medical School takes it a step further. He believes the reason for daydreaming is indirect – that daydreaming creates memories for events that haven't happened in order that we can then use those memories if we need to. Anyone who goes on a plane wonders what might happen if it were to crash. Bar's idea is that if a plane did actually crash, the memories of all those daydreams from previous flights would come into play and might help save you.[101]

The evidence is mounting to suggest that our minds are skewed towards thinking about the future. In one study people had to imagine themselves living either in the present, 10 years ago or 10 years hence, and then decide as fast as possible whether a list of events on different dates would then have occurred in the past or in the future. They were much faster when it came to events in the future, even, and this is where it gets interesting, if they were imagining them-selves living 10 years ago.[102] This suggests we constantly lean towards future thinking. Add to this the strength of emotions connected with future thinking and the finding that it appears to be the brain's default mode when unoccupied, and it seems clear to me that the future is the most dominant time-frame when it comes to our experience of time. Our inclina-tion is to pitch ourselves forward in time. We take our ability to imagine the future for granted, barely even considering it to be a skill, yet this imaginative construction process has been described as being near the 'apex of human intellectual abilities'.[103] It is our ability to time-travel mentally that gives us the experience of mental reality. It roots us.

AN ERRONEOUS FUTURE

There is just one problem. We may be forward thinking, but that doesn't mean we are good at predicting the future or imagining it objectively. The future is a time-frame we find difficult to grasp accurately. In his extensive research, the American psychologist Dan Gilbert has discovered that we make various types of errors when contemplating the future. The first is caused by the way we meld memories from the past to imagine the future. This memory remix allows us extensive imagination, but it causes us to base our ideas of the future on the past without any evidence that it will be the same. If you know you are due to go to hospital you will think back to the last time you visited a hospital and assume that it will be similar, even if the previous visit took place in a different hospital in a different town a decade earlier. Financial experts warn that previous performance of an investment fund should not be taken as an indication of future performance, yet how many people would really take this information at face value and deliberately seek to invest in a fund that has performed badly in the past? The nature of memory skews our thoughts. Our cognitive processes favour the extreme, the first and the most recent. So when we imagine the future it is examples of these kinds of events which come to mind, while the typical is ignored.[104]

There's a second problem. When we simulate a future event in our minds we tend only to consider the chief features, the parts we consider integral to the experience. So if you're heading out of a city for a country walk and a pub lunch you might imagine yourself crossing stiles, walking along lanes wondering what it's like to live in a pretty cottage like that, climbing a hill, descending into a green valley and stopping for lunch in a cosy pub in the village. You might be right. It might be just like this. But the outing will also include some less attractive parts, which tend to be omitted from your advance picture of the day – queuing in traffic to get out of the city, stopping for petrol, searching for somewhere to park, getting lost on the walk perhaps, and then arriving at the pub and discovering there are no free tables. This focus on what we consider to be the chief features of a possible future event leads us to consider only the best bits. For a negative event we do the opposite. We dread all of it, focusing on the bad bits, when some of it will be fine. Visiting the doctor for a physical examination might not be very pleasant, but neither is every part of the visit *un*pleasant. Some of it is neutral; reading a magazine in the waiting room, hanging up your coat, chatting with the doctor, arranging another appointment with the receptionist. As a proportion of the whole experience the duration of the examination might be short, yet beforehand it's the only part you envisage, causing you to overestimate your emotional reactions. You could argue that at least nothing's as bad as we fear, that we'll be pleasantly surprised later on, but the same phenomenon can lead us to make some strange, even wrong, decisions. This is known as the 'Impact Bias'.

We expect the best of good events and the worst of the bad. We imagine that if something grave happens to us, we won't be able to cope, and that if something positive happens it will make us so happy that our lives will be transformed. But in both cases we will still be the same people we are now. After the initial phase, good or bad, our emotions will subside and we will feel only a little better or worse than now. This is because of the way we view time in the future – to consider an event in real time would take too long, so we truncate it, imagining more of the early moments. If you are moving in with a partner you picture the fun of the first year living together and sorting out the flat, not the more routine life of the fifth or tenth year. Daniel Gilbert has researched people's ideas about what life would be like if they had the glorious elation of winning the lottery or the appalling experience of becoming paralysed.[105] People imagine that if they won the lottery they'd have the elation of champagne celebrations, posing with a giant cardboard cheque for millions, test-driving sports cars and taking all their friends on holiday. Non-stop fun. In the case of disability they imagine the shattering shock, the loss of their job, the work of converting their home. Everything ruined. When imagining either situation people tend to focus on the initial impact, yet assume these feelings will be long-lasting. They forget that they will adjust. Some of these initial sensations of either joy or despair will wear off. If you get that longed-for promotion, researchers have found that the extra happiness it brings lasts for only about three months. After that you become accustomed to your new life and have experienced some of the disadvantages of the new job as well as realising that many of the previous irritants are still there. You still commute. You still have to get up early. You still have one annoying colleague. Likewise if you are forced to change jobs and leave the one you loved, then after a while you adjust to that too. Gilbert has found that even with a serious disability, although the transition period might be devastating, in the mid- to long-term most people cope better than they had anticipated. They end up not far below their original happiness levels and if they were fairly happy in the first place they might well still end up a lot happier than the lottery winner who was less happy to start off with and whose joy gradually wears off.

There is a whole list of real-life examples where people have overestimated their feelings in the future. People moving from the mid-west of the USA to California predicted that they would be happier in their new homes, believing the sunny weather would transform their lives. Sadly it didn't; the weather is just one factor contributing to well-being in life. When another group of people had just received good news – test results revealing that they did not have HIV as they'd feared – they didn't feel as elated as they'd expected.[106]

Dan Gilbert and his colleagues invent hypothetical scenarios and get people to imagine how they would feel if they happened. Sometimes the situations are everyday – their team wins or loses a game of baseball – sometimes they are more serious. Gilbert asks a mother of two to imagine how she might feel in seven years' time if one of her children had died now. She predicts that will feel dreadful all the time, forgetting that although her experience would be horrendous, and although life will never be the same again, there would be some moments of joy with her other child.[107]

Curiously these studies have occasionally forecast real events. In the year 2000 people were asked to imagine their emotions on hearing the news that the Space Shuttle Columbia had been destroyed, killing a dozen astronauts.[108] In their version it crashed into the Mir space station, but it was another three years until the space shuttle really did explode, killing all seven crew members. The same study asked them to envisage the USA toppling Saddam Hussein in a second Gulf War. Again they were three years ahead of the real event.

To sum up so far, future thinking is crucial and might even be the brain's default position, but our judgements are skewed by a tendency to concentrate on the initial and chief features of an event and to base our predictions on our most extreme past experiences, rather than the more typical. And just like the children with the pretzels, who found it hard to imagine that they could possibly feel anything but thirsty the following day, even as adults we find it hard to discount the way we feel in the present. When people are not hungry they say they're not keen on the idea of spaghetti Bolognese for breakfast, but pose the same question when they do feel hungry and suddenly the idea of an evening meal for breakfast becomes more appealing. The mind can generate very convincing simulations of the future, but they are not perfect, especially when it comes to their emotional content. We are simply not very good at predicting how we'll feel in the future, which can lead to some unfortunate decisions.

BAD CHOICES

The way we hold the future in mind has important consequences for decision-making. The Impact Bias affects the choices we make, as can errors in our predictions about what makes us happy. If we decide to move house we become convinced that our future happiness depends on finding the right home, in the right location. In fact our happiness in that house will be far more dependent on the relationship we have with our partner or housemates.[109] Likewise if someone announces they are leaving their current job for another that pays a bit better, most of their friends will consider it to be a logical move, despite the fact that our well-being is influenced more by our colleagues and the atmosphere at work, than by a small pay rise.

> You have two projects to complete – one is easy because it's in
> English, but the topic is the history of social psychology, some-
> thing which doesn't especially appeal to you (although I can't
> think why not – it's interesting, genuinely). The second project
> is harder because it is written in French, but it concerns romantic
> love, so you might even learn something useful. One project is
> due in next week, the other in two months' time. You can choose
> which project to do first, but whichever you select you will only
> receive the instructions one week before the deadline. Which
> would you do first?

When this study was conducted with students in Israel, with the projects either in Hebrew (the easy option) or in English (the difficult option), the overwhelming preference was to do the easy but dull project first and to save the interesting but trickier option for later.[110] When they contemplated the future, people didn't seem to worry about how long the project might take. They were convinced that in the future they would have more time, so it wouldn't matter. I've already discussed how our optimism increases the further into the future we transport ourselves and nowhere is this more evident than with time itself. Despite all the evidence from past experience, we are always certain that in the future we will have more spare time.[111]

Students were given two lecture options for the following year. They could attend an interesting lecture on the other side of town or a boring lecture in the hall next door. Most opted for the interesting lecture. No surprise there perhaps.

But if they were told the lecture was tomorrow instead of next year they reversed their decision. Once they took the practicalities into account they realised they had too much to do to travel across town and so they chose the dull, but convenient, lecture instead.[112] Although we know intellectually that every activity we choose comes at the expense of other ways we might use those hours, this only seems to matter to us in the near future; in the distant future we simplify the situation and omit crucial elements, forgetting that in a year's time we will be just as busy.

This optimism that we have more time in the future is fascinating, because we never seem to learn that it's not true. So we postpone going to the gym today because we're too busy, but sincerely intend to go tomorrow. We retain a constant optimism about our future selves. We'll be better. We'll be more organised. And therefore we'll have more time. A year from now we picture ourselves as consistent, methodical people who could easily fit in some extra activities. But when we consider ourselves next week, we know we couldn't possibly take anything else on. In the near future we take into account the circumstances that might thwart us – but the person we see in the distant future appears unaffected by anything as ordinary as a broken-down train. We even use simpler adjectives to describe ourselves in the future.[113] This rosy view of the future causes us to try repeatedly to cram more into a week than is possible. If someone asks me to give a talk in Wales some time next year, I might say yes, thinking that I'll arrange my work so that I have a free day to take the three-hour train journey to Wales. It seems like a good idea until the date approaches and my diary is so full that I wish I'd never agreed to it. Yet if someone asked me if I could go to Wales to give a talk next week I know immediately that I must decline. This optimism about free time in the future can lead to procrastination.

It is often assumed that procrastination is simply caused by laziness and a lack of focus. In fact we sincerely believe that in the future, even next week, we will have fewer demands on our time. Tasks need not be tedious for us to procrastinate. Companies are pleased to offer hugely discounted online vouchers to use in the future because they know that even when the voucher is for something as pleasurable as a good meal, the chances are that many of us will never get round to spending it. Suzanne Shu demonstrated this in a study where people claimed to prefer vouchers with distant deadlines, but were in reality much less likely to spend them than vouchers that expired in two weeks.[114] She also found that visitors to cities do more sightseeing in three weeks than those living there do in three years – because they have a deadline. When there's no deadline people don't see the sights because they continue to believe that they'll have more free time to do so at a later date. We've all done it. For nearly a decade I had the chance, as a result of my partner's job, to go and see Prime Minister's Questions from the press gallery at the House of Commons, but somehow I was always busy on a Wednesday. It wasn't until the very last Wednesday that a pass was available to me that I finally went, even though it was something I very much wanted to do. In a study in Chicago, Shu even discovered that long-term residents who were moving away for good were trying to

intersperse their packing with hurried sightseeing because
they had never seen their own city.

FIVE YEARS TO REACH THE WORD 'ANT'

In 1857 the Philological Society of London made an
announcement – the formation of the Unregistered Words
Committee, which would collect all the English words
currently absent from the available dictionaries. Five
months later the Dean of Westminster Richard Chevenix
Trench went a step further when he delivered a two-part
paper calling for a complete re-examination of the history
of the English language from Anglo-Saxon times onwards.
This would be the finest dictionary every created. By 1860
the plans were in place and it was confidently announced
that the dictionary would be ready within two years. It's
fair to say that there were some unavoidable delays. A
young man named Herbert Coleridge had begun work on
the dictionary, starting with words beginning with A–D,
but was taken ill with tuberculosis, reputedly exacerbated
by sitting in damp clothes at a meeting of the Philological
Society itself.[115] Just two weeks after presenting the society
with the first set of words, he died. After that things moved
slowly. In 1879 a deal was struck to publish the dictionary
with the Oxford University Press, with a new completion
date set for 10 years later. But after five years they had
only got as far as the word 'ant'. Nobody had anticipated
the work involved in tracing the history of words across
seven centuries while keeping up to date with a language that
was constantly evolving. Decades of research followed until

finally in 1928 the complete *Oxford English Dictionary* was finished. Immediately considered out of date, revisions began at once.

Even compared with the over-runs which blight government procurement of new computer systems or public buildings, a prediction of two years for a project that took 71 is some underestimate. Yet looking from the outside it seems obvious that the task was so substantial that even the longer 10-year deadline was over-optimistic. We have the advantage of hindsight of course, but our position as outsiders brings us a second advantage – an unexpected degree of skill at predicting how long someone else's project will take to complete. When a friend tells us of their disappointment that their kitchen is still not ready despite the builders' promises at the start, we are not at all surprised. Yet this skill deserts us when it comes to our own projects. This tendency to underestimate how long a task will take is called the 'Planning Fallacy'. The cause rests once again in the key feature of future thinking that I've already discussed – the lack of detail. The further into the future we look, the more we ignore the details, but, and this is where it becomes more curious, we *do* consider the details when contemplating someone else's future. When reflecting on another person's project we consider both the length of time similar tasks have taken them and the factors that might disrupt them – illness, unexpected visits from friends, tiredness, etc. When estimating how long our own project will take us we ignore all this information and focus only on the features of this singular task.[116] The beauty of the study which best demonstrated this was that for once it didn't use hypothetical situations, where you can never be certain that this is how people would behave in real life. Instead they tested students who were trying to finish their theses. They found that they were much better at predicting how long *other* students' theses would take. When it came to their own they did on occasion refer to past projects, but not to enable them to make a more accurate prediction, rather to justify their optimism. They seemed to forget all the occasions in the past where their good intentions had been disrupted by unexpected events.

There are many methods of avoiding the Planning Fallacy and making more accurate predictions about the demands of a task. I'll discuss this some more in the next chapter, but here are two simple techniques to be getting on with. Either ask someone else to assess how long your project will take or, if you want to decide for yourself, then adopt the strategies that they would use and apply them to your own situation. Deliberately think back to all similar past occasions, but compare them with the present circumstances before you make your estimate. Research has shown that just as our idealised future tells us nothing, neither does considering the past alone. If you really want to know how long a job will take, you need to think about past occasions and then look at the details of the new task to see how it matches up. Then add on some time for the types of disruptions you have encountered in the past and a little more for the fact that unfortunately you are not suddenly going to be transformed into a super-organised person who doesn't need to sleep.

I've discussed some of the errors we make in future

thinking, but there are two final aspects to the way we consider time in the future that I want to think about. First, some people spend more time considering the future than others, which brings us to the topic of time orientation.

ONE MARSHMALLOW OR TWO?

If I offered you the choice between eating one marshmallow now or two if you are prepared to sit and wait with the marshmallows for 10 minutes before you eat them, which would you choose? It will of course depend on your liking for marshmallows and whether you have better things to do with the next 10 minutes. You might decide that it would be easier to take one marshmallow now, knowing you can buy yourself a whole bagful on the way home if it gives you a taste for them. But four-year-olds don't have this option, and so the marshmallow question is one they take very seriously. Not only that, but the decision they make can predict how well they will do at college or the likelihood that in 20 years' time they will take drugs.

The marshmallow studies are some of the most famous experiments to have been conducted at the Bing Nursery, found at a crossroad on one side of the Stanford University campus. When university staff send their children to this nursery, part of the deal is that they give permission for them to take part in psychological research. It's easy to see why they say yes. The nursery is packed with toys, games, craft materials and happy staff. The sunny California weather means the children are free to wander out into the large landscaped play area whenever they feel like it. But despite all these facilities, there is one part of the day that many of the children like better than any other – the moment when a researcher invites them into one of the special rooms surrounding the central courtyard. These rooms are small and contain nothing but a table, some child-sized chairs and a video camera. At first sight, playing outside on the climbing frames would seem like a much better prospect. The children don't know they are taking part in research which could transform our views on child development and influence policy on childcare. They don't know that these studies could have a lifetime of implications. What they do know is that for a short while in that room they and they alone will have the focused attention of an adult who will give them a new game to play. When I visited it was clear that this was a rather special nursery, one that has been home to more discoveries in developmental psychology than any other in the world.

The psychologist Walter Mischel began his marshmallow studies here in 1968. Like many classic psychology experiments it wouldn't be allowed today, not because of the tightening up of the ethics of experimentation, but because – the staff told me – today's parents would find it unacceptable that their children were given sweets at nursery, even if it were just one marshmallow or two.

The study works like this. A child sits at a table on which there are two white plates and a small hand bell. On the centre of one plate sits a single, pink marshmallow. On the other plate there are two. The experimenter says she is leaving the room and that the child has a choice. If she wants two marshmallows she needs to wait a while until the experimenter returns. Alternatively the child can choose to ring the bell and will then be allowed to eat the marshmallow immediately, but she will only get the one. It is a straightforward choice – one sweet now versus two in 10 minutes' time.

Between 1968 and 1974 more than 500 children took the marshmallow test. While they waited there was nothing to play with and nothing else to do apart from gaze longingly at those plump, pink marshmallows. This test measures a child's ability to resist temptation, to delay gratification. The early studies weren't filmed, but watching later films of replications you can see the tactics employed by some children to distract themselves from the marshmallows. A few cover their eyes. Others sit on their hands. Some stare determinedly at the ceiling. They do anything they can to stop themselves from focusing on those sweet, chewy marshmallows. Not a single child asks why they should have to wait in order to get two sweets; they seem to take the rules at face value. As Walter Mischel told me, 'To them this is real life.'

It would appear to be a simple test of patience and self-control, but when Mischel followed the children up many years later he discovered his test was far more powerful than he would ever have guessed. The children who had waited patiently for the reward of two marshmallows were more likely to have succeeded at school, at college and at work. Those who had gone straight for the single marshmallow were more likely to have taken drugs, to have lower incomes or to be in prison.[117] I spoke to Carolyn Weisz, who was one of the children who took part in Mischel's marshmallow test 40 years ago. She remembers the nursery but doesn't know whether she took the marshmallow or not, and because the study still continues all these decades later, the researchers are careful never to reveal an individual's results. Today they are sending laptops out across the country for these now middle-aged participants to complete batteries of tests. Coincidentally Weisz is now a professor of psychology herself and remembers her time at the Bing Nursery fondly.

The purpose of this study was never to label children, and of course the results are averaged over large numbers so if you were to try the test on your own child and found them to be impulsive, they are not doomed to a life of crime. Experiments are now taking place to see how you can teach the most impulsive children to employ those same distraction strategies used by the children with the best self-control.

So where does time come into all this? The marshmallow studies tend to be characterised as tests of impulsivity, but they could also be useful in measuring the extent to which the children consider the future – their future orientation. When Mischel studied teenagers, he found that those who were good at delaying gratification were also judged by their parents to be better at planning than the adolescents who had opted for the single marshmallow a decade earlier. In a more recent study where teenagers were asked to choose between a small amount of money immediately versus a larger amount later, personality tests showed it wasn't their impulsivity that best predicted their answers, but the degree to which their thinking was orientated towards the future. They found that 10- to 13-year-olds were prepared to accept less money if they could have it sooner, while over-sixteens chose to wait for a larger sum.[118] This suggests that a quantifiable change in thinking about the future occurs during the middle-to-late teens.

Adolescence is known to be a stage of life where people take risks without considering the consequences for the future. The response from those designing health promotion campaigns has been to focus on the short-term – anti-smoking campaigns featuring jars of make-up turned into ash trays to emphasise the effect smoking can have on your skin today, instead of concentrating on the impact on your lungs in half a century's time. Various physiological explanations have been sought for this teenage tendency to take risks, including the slow development of the frontal cortex, the influence of hormones on the immune system and the continuing development of working memory. The cognitive networks that bring control take longer to mature, so in the meantime emotional networks and reward-seeking mechanisms take over in the brain and the result is that the teenager takes risks. But future orientation seems to be crucial too. Many studies have found that the older adolescents are, the more likely they are to plan and to take future considerations into account when making decisions.

FUTURE-ORIENTATED THINKING

The charismatic psychologist Philip Zimbardo (who thought nothing of giving an undergraduate lecture wearing a glittery top hat) is famous for his Stanford Prison Experiment, in which he transformed the basement of the psychology department at Stanford University in California into a make-shift jail, divided volunteers into guards and prisoners and then sat back to see what happened. It didn't go well. Or rather it didn't go well for the prisoners, who soon found themselves shockingly mistreated by the guards. It went better for Zimbardo, securing a place for his study as a classic in social psychology. In the end his research assistant – who later became his wife – convinced him that the violence was getting too serious and he must call a halt to the study. Compared with that experiment, his more recent research on time orientation seems quite tame, but is significant nonetheless.

Individuals vary in the degree to which their thinking is orientated towards the future. Inevitably circumstances can sometimes prevent us from planning ahead; if you don't have anywhere to sleep next week, you are unlikely to focus on your career prospects five years hence. But while future thinking could be the default when the mind is unoccupied, some people actively choose to reflect on the future far more than others. Charles Darwin revealed himself to have an extreme future-orientation in his thinking (perhaps because he was concerned with looking so far into the distant past) when he wrote a list of the pros and cons of getting married. A con of not marrying was the lack of anyone to care for him in his old age. Another was the lack of children, after which he put in brackets 'no second life'. I think it's safe to say that Darwin was looking well into the future.

The Zimbardo Time Perspective Inventory assesses these temporal biases and divides them into six time perspectives: past-negative, past-positive, present-fatalistic, present-hedonistic, future and transcendental future. Using his questionnaire you can measure the extent to which your outlook falls into these categories. Naturally none of us spends our mental lives in one time frame alone, but for most people, two or three of these six time-frames predominate. Not surprisingly those with a present-hedonistic orientation live in the moment and are more likely to gamble, drink a lot and take risks on the roads, while those with a future orientation (those same teenagers prepared to wait longer in exchange for a greater sum of money) do better in exams and are more likely to floss their teeth.

So which time orientation brings maximum happiness? The past-negative doesn't sound like a good place to be, but adopting the modern view of nostalgia Zimbardo views the past-positive orientation as a good thing, provided you're not mourning the loss of your past. Many researchers have found people with a future orientation to be generally happier, but Zimbardo disagrees, warning that an excessive future orientation can lead to workaholism, a lack of social contact and a poor sense of community. His research shows that we should aim not for one particular orientation to dominate, but for a balance between them. This seems to be easier said than done. His colleague's study found that only 8 per cent of participants had a fully balanced time perspective. Curious considering he believes this balance is required for happiness and yet in well-being surveys a far higher percentage of people than this describe themselves as being satisfied or very satisfied with life.

Zimbardo does suggest that when you've worked out which time orientation might predominate in your life by completing his online questionnaire, then making a few very small changes to your life could make a difference. If you are low on past-positive, you could choose to phone an old friend to reconnect, you with the past. If you're low on present orientation, then set aside an hour to do something utterly absorbing. To increase your future-orientation, plan a future event in detail. The greatest chance of happiness, he says, comes from a combination of a past-positive and future perspectives, with just the right amount of present-hedonistic, living-in-the-moment thrown in. Examining your time perspective through surveys like Zimbardo's can be an illuminating exercise, but making lasting changes could be difficult.

LOOKING BACK, LOOKING FORWARD

There is one other element of our perception of the future that varies among individuals and societies. This is not about whether you focus on the past or the future, but how *far* you look ahead or behind you. The first time capsule from the long-running British children's TV programme *Blue Peter* was buried in 1971 before being moved to a new underground site where it was joined by a second capsule in 1984. The capsules remained under the Blue Peter garden and, despite the rumour that the BBC had lost the map with their location, were unearthed during a special programme in the year 2000. The capsule had only been under the ground for 29 years, which in the scheme of things doesn't seem that long. Even the ex-presenter Peter Purves, who returned for the opening of the box, admitted he was surprised so many people were interested. But they were. The presenters waved sheaves of letters from the people who had, as requested back in 1971, written in to remind them to dig up the capsule. Sadly when the current and ex-presenters gathered for the solemn opening of the box, they had to obscure their disappointment. The film of the box's burial 29 years earlier had talked of the lead lining and the bolts sealing the box 'absolutely tight'. Sadly not tight enough. Water had seeped in and removal of the lid revealed something of a muddy, sodden mess, although the first set of decimal coins survived, as did the Blue Peter badges. Other capsules won't be opened until 2029 and 2050. This time-frame seems reasonable for a TV programme hoping to evoke a little nostalgia and create moments in its own history, but it is dwarfed by the many time capsules buried in Japan with instructions to keep them sealed for 5,000 years. Now that is what I call temporal depth!

So what is it that dictates our temporal depth? Winston Churchill once said, 'The longer you can look back, the farther you can look forward.' Recent experiments have proved him right. Here's a test:

> Think of an event that happened to you a short time ago and write down the date, then write down an event and date that happened a middling time ago and finally an event which happened a long time ago and its date. Now write down three more events and their corresponding dates, but for occasions that you think you might experience in the near future, the middling future and the distant future. Compare the dates. Which are more distant, the dates of events in the past or events in the future?

As usual there are no right or wrong answers, but most people choose dates in the future which are farther away from the present than the dates they select for the past. The organisational psychologist Allen Bluedorn found that most people see the short-term future as being five times longer than the short-term or recent past. Nearly everyone defines the short-term past as falling within the last six months and two-thirds say it means an event happened as recently as a fortnight ago or less. This suggests we look slightly further into the future than into the past. And this next finding is where Churchill was onto something: the more distant the dates people selected for the past events, the further ahead they looked for the future events. So it is true that the further you look back, the further you look forward (reminiscent of Darwin and his marriage list).

Even the order in which you contemplate the past and the future can make a subtle difference to your thinking, and one you can utilise too. A group of CEOs at Silicon Valley companies were asked to name 10 future events followed by 10 events from the past. A second group did the same except they were told to list the past events first and then the future events. Whether the future was considered first or second made no difference to people's temporal depth, but if they considered the past first, they chose future events an average of five years further away.[119] This difference is striking. It provides a practical lesson for those in business: the older an organisation, the further its members tend to look into the future. There is a management exercise that attempts to harness this idea by instructing managers to say what their hopes are for their company in the future-perfect tense rather than the simple future. So instead of saying, 'We will sort out the problems on the production line,' they say, 'We will have sorted out the problems on the production line.'[120] The idea, paradoxically, is that this tense feels closer to the past tense and so it makes it easier to imagine future possibilities. Using the past to imagine the future can even make a difference to the quality of imagination. If people are asked to describe an imaginary car accident which happened in the past tense, they do so in far more detail than if they're asked to describe an equally imaginary accident in the future tense. Both events are fictional. Both require imagination. And yet one is easier.

This illustrates once again how the mind constructs its own sense of time in the future just as it does in the past and the present. Our concept of the future is tied up closely with our perception of the past. By now it should be clear that future thinking has a powerful effect on our actions. It gives us foresight and imagination, and allows us to formulate plans, but it can also warp our thinking, causing us to make decisions we might regret, from the marshmallow-like trivial to the fatally serious. But we can use future thinking to our advantage, along with the other time-frames, and it is the harnessing of all this research on time perception which I shall discuss in the final chapter.

CHANGING YOUR
RELATIONSHIP WITH TIME 6

WE HAVE TRAVELLED mentally backwards in time and forwards in time. We have seen how our minds actively construct our experience of time via memory, attention and emotion. Although no dedicated organ to measure time has been found in the brain, we are able to assess its passing. For the most part time feels as though it moves smoothly, yet we are repeatedly surprised by the tricks it seems to play on us. Our relationship with time is not straightforward, which is what makes it so fascinating.

In this book I've scoured the literature for what I believe are the most informative studies on time, conducted by researchers from all over the world. The question now is how to put this knowledge about time perception and mental time-travel into practice. This is not a self-help book, but a sweep across the research in this field does indicate certain ways in which we can, if we choose to, both harness and mould the way our minds perceive time. Every recommendation I'll make in this chapter is evidence-based; none of this is advice I've simply made up. That would be a waste

of your time. In many ways research in this area is just
beginning, and I have no doubt that as the field of the
psychology of time begins to expand there will be new
insights for us in the future. But taking the best evidence
we have to date, there is plenty we can already put into
practice.

If life feels as though it is racing by, with each year
passing faster than the last, then this is the chapter where
you can discover how to slow time down, although I will
question whether this is something you really want to do.
You will learn to date past events more accurately and to
use your time wisely. But remember that we don't all have
the same problems with time. Some feel as though it's
accelerating; for others the hours seem long. Some find
their minds wandering to the future and worrying obses-
sively; others forget what they were intending to do. In this
chapter I will deal with eight different problems.

Reading this book you will probably have already begun
to think about how you personally view time. Do you see
it laid out in front of you in imaginary space? Would you
put the future on the left or the right? When you considered
Thomas Cottle's studies with US Navy personnel perhaps
you drew your own circles of time on a piece of paper. If
so then you might have sensed already that one time-frame
is more salient for you. Or perhaps you went online to fill
in Zimbardo's Time Perspective Inventory.

Do you move along a time-line towards the future or do
you sit still while the future comes towards you? Think back
to the question about moving Wednesday's meeting forward
two days. If you think it's now on Friday then you see

yourself as actively moving along through time. If you think it's on Monday then you stay still and it's time that moves towards you. Each of these tests begins to explain how you have personally created your own perception of time. This is the impression of time you have constructed for your mental world, but because you actively create mind time, you can also influence it. Now take your pick from time's challenges.

PROBLEM 1: TIME IS SPEEDING UP

While I've been writing this book, the question that the greatest number of people have asked me is how they can slow time down. As I've discussed, it is very common to have the sensation as you get older that time is speeding up, that the years are flashing by. It has been found that if you ask people to guess when three minutes have passed and let them count the seconds off in their heads in one-crocodile, two-crocodile, three-crocodile-fashion, young people do very well, over-estimating by an average of only three seconds. Middle-aged people over-estimate by 16 seconds. But 60- to 70-year-olds overestimate by 40 seconds, which is a lot on a three-minute duration. It is as though their internal clock has slowed down so more time passes than they expect, which gives them the sensation of time speeding up.[121]

The perception feels very real. The question remains, what to do about it? But before we approach that I would like to pose one more question: do you really want to slow time down?

If you think back to the research on estimation of the passage of time, there were various circumstances where time warped to the extent that it felt protracted – Mrs Hoagland in bed with a temperature; Michel Siffre in his ice cave, lying on his damp camp bed, surrounded by rotting food, longing for dry socks while he gradually went colour-blind; people in such despair they are contemplating suicide, where time expands until one hour feels like three; Alan Johnston counting off the hours during each long night in his cell, fearful for his life. For all these people time felt slow (even though Siffre later discovered it had been passing much faster than he thought). Is this really something we want to simulate? Boredom, anxiety and unhappiness will all slow down time, but none of them are very appealing mental states. I would argue that if you live a life where time goes fast, this is a sign of a life that is full and prob-ably fulfilling, not empty. Slow-moving time might be less desirable than you think, unless of course you could find a way to isolate the more pleasurable experiences and make them linger.

There have been deliberate attempts to lengthen the experience of time using hypnosis. Back in the 1940s two American psychiatrists Linn Cooper and Milton Erickson hypnotised some volunteers. While they were in a trance they were instructed to picture themselves going for a 10-minute walk, but were given only 10 seconds in which to imagine the entire thing. Once they were out of the trance they were able to describe in detail a walk that would have taken 10 minutes. The question is whether they had actually learned to distort their perception of time, slowing a minute down to 10, or whether they simply had particularly good imaginative skills? Decades later the psychologist Philip Zimbardo also attempted to distort time through hypnosis. Knowing himself to have an especially strong orientation to the future and a reluctance to enjoy himself in the present, he arranged for a colleague to put him into a trance and then suggest to him that he allow the present to expand and fill his mind and his body. Zimbardo believes that for him it worked; he began to notice the smells surrounding him and the extraordinary colours in a painting on the wall.

So what if you don't want to resort to either hypnosis or suffering to make time slow down, but instead just want to lose that unsettling feeling that every week is slightly more fleeting than the one that's just gone before, or that it's nearly Christmas yet again. There is a way of stopping the years rushing by and for this you need to harness the Holiday Paradox (the feeling you get that your holiday passes in an enjoyable flash, but as soon as you get back it feels as though you were away ages). To recreate the holiday feeling, some people go to the trouble of moving their entire lives to a holiday resort. In her ethnography of the British expatriate community in Spain, the sociologist Karen O'Reilly found that part of the appeal of a new life abroad was the desire to live more in the present.[122] The British people she interviewed on the Costa del Sol liked the fact that their new friends knew very little about their past and that it was rare for anyone to discuss the future. People told her they had deliberately only planned one element of their futures – that they never wanted to return

to live in the UK. Other than that, she found that few
people had any plans beyond the next day. They had been
so successful in ceasing to live by the clock that it made
her research rather challenging. She would arrive on time,
only to find her interviewees walking down the drive with
towels under their arms, off out for a swim – and surprised
that she couldn't go with them. When she got lost one day,
arriving over an hour and a half after her appointment time
to interview another couple, they hadn't even noticed she
was late and found her apologies amusing. What's striking
about the British community on the Costa del Sol is their
active decision to emigrate in search of a life that both has
a slower pace and is spent more in the present. They are
attempting to harness the Holiday Paradox, to create long,
lingering days on which to look back. The problem is that
the retrospective impression of these days as long relies on
novelty, and although life abroad might have fewer routines
than at home, the new memories which make time seem
lengthy in retrospect will inevitably get rarer as people
become accustomed to their new lives abroad. O'Reilly even
suggests that they were attempting to challenge the idea
that time flows in one direction, trying to somehow stall its
onward march.

To slow down the passing years at home we need to
recreate the Holiday Paradox by studying the features that
make a holiday unusual. First, they involve few routines.
But routines are hard to avoid in everyday life; repetitive
chores such as cleaning are always going to be there, and
if you have young children who need routine, you can't
abolish it altogether. What you can do is to try to add variety

wherever possible. If you can create a life which feels both novel and entertaining in the present, the weeks and years will feel long in retrospect. If you have any choice about your route to work then keep varying it, even if the alternative takes a few minutes longer. This can prevent the auto-pilot effect where everything is so familiar that you arrive at work and can't even remember some parts of the journey. As soon as you vary the routine, you are forced to be mindful. You notice more things around you and this novelty tricks you into experiencing that time as slightly longer in retrospect. Now you might not want to do this every day. One of the reasons for always taking the same route to work is *not* to think about it, to feel you're giving your brain a rest. So you might not want a daily adventure, but you can decide to look for something different each day. What colour are most people wearing on the bus? Which building has the nicest roof?

On holiday you constantly have new experiences that create brand new memories, and – looking back – this is what gives you the sense that you have been away for ages. So the more memories you can create for yourself in everyday life, the less the weeks will rush by. If governments are serious about increasing well-being they could even encourage employers to create more variety at work with a lunchtime talk, a job swap for a day or allowing people to carry out their tasks in a different order, or from a different location. If you fill your weekend with activities and go out for the day on both Saturday and Sunday to do something new, the minutes and hours will pass fast because you are so absorbed, but at the end of the weekend you will feel as though you have had more than two days off work. If you did something different every weekend, you would make so many new memories in a month that the weeks would cease to rush by. Modern research bears out the advice from the philosopher Jean-Marie Guyau back in 1885. He said to lengthen time, 'Fill it, if you have the chance, with a thousand new things.'

Now, you do need to have a lot of energy to pack your weekends with different activities, and if you've had a hard week at work you may well not yearn for a host of new adventures, but rather for an empty weekend. A weekend spent at home, reading the papers, tidying up, watching TV and phoning a couple of people will relax you, but it gives rise to few new memories and soon that weekend will not stand out from any other, making time appear to have gone faster. So there is a trade-off here: do you want to slow time down or spend your spare time restfully?

Disappointingly, watching TV isn't the answer. When you are tired and want to do nothing it may seem like the perfect activity; you don't have to move, you don't have to concentrate too hard and yet it simultaneously distracts you from your own worries while entertaining you. No wonder TV is so popular. But the problem with television, and the same applies to computer games and time spent online, is that it doesn't lead to the formation of as many memories as non-screen activities. There will of course be exceptions, in programmes so powerful you never forget them. I'm convinced I have created plenty of memories from watching the drama *The Wire*, but I'd still have to admit that the memories from all five series probably don't add up to 45 hours' worth of another, more energetic, activity. Matt, who plays computer games early in the morning and late at night, tells me that he doesn't remember whole games. While he waits in the virtual lobby for 12 people around the world to get ready to play shoot-em-up games, time drags, but once he is playing his absorption is so intense that time appears to shrink. This is exactly what researchers found when they studied gamers in a video gaming centre in Quebec City.[123] Afterwards, people underestimate the duration of the game. Yet in terms of memories Matt says he only remembers the highs (the killstreaks where lots of people die), the lows (where he dies) and any new techniques he has learned (if your gun is large enough you can apparently shoot through walls).

I am not saying that you should never watch TV, play computer games or spend a weekend doing very little. But if you really want to stop time speeding up, the answer is to devise an energetic timetable and only to watch TV when you know it will be memorable. Armed with this knowledge, it's for you to decide which is more important to you. You might choose to spend less time in front of screens and to fill your time with memorable activities. This will give you copious memories, creating the impression that lots of time must have passed. Time will slow down. But perhaps you don't want to do lots of new activities. Maybe part of growing older is having the option to spend more hours doing what you know you like best, rather than seeking out new experiences. Why learn to sail if you know you hate every other water sport you have ever tried? Why endlessly seek new restaurants if you live two minutes away from the

place that always serves a meal you really like? The choice
is yours. Once you know *why* you have the sensation that
time is speeding up, it might matter to you less than you
think. Or you might decide that since time rushing by is a
sign of a busy, happy life, that it doesn't matter enough to
sacrifice resting or watching programmes you enjoy. As
Pliny the Younger wrote in AD 105, 'the happier the time,
the shorter it seems'.

PROBLEM 2: MAKING TIME GO FASTER

Time has such a powerful hold over us that we both hate
and fear wasting it. On days when I make radio programmes
I have no spare time. Yet the fear that I might finish writing
the script early and have nothing to do for an hour while
I wait to go to the studio is so strong that I constantly email
extra work to myself or carry extra reading with me, just
in case.

This is one of the reasons we like to queue, to be certain
that we don't have to wait a moment longer than is fair.
Many cultures regard it as democratic that they stand in
line regardless of rank (with the exception of the advantages
bought with a business-class airline ticket, of course). Barry
Schwartz, a psychologist who has studied queuing, believes
the reason we hate it if someone else pushes in is that we
have resisted our own urge to push in and feel others ought
to do the same. We know we all have to be saved from our
worse selves.[124]

We judge the time we spend waiting to be longer than
it actually is because we're in anticipatory mode. Yet if

someone offered you a 10-minute rest doing nothing (which is effectively what queuing offers) in the middle of a busy day at work, you would probably welcome it. When waiting is forced on us in a queue we find it hard to savour the experience as the bonus of some time to do nothing. Expectations, experience and culture all influence our tolerance for queues. Having spent her young adult life in communist Poland, the writer Eva Hoffman says that because there was nothing to hurry for, queues did not present a problem. But after living in the United States and returning to Eastern Europe after the fall of communism in 1989, she found the queues intolerable.[125]

Sometimes we need to find a way to make time speed up, whether in a trivial situation like a post office queue or in grave circumstances like Alan Johnston's. He told me that when he relates his experience to people (which is rare because, extraordinarily, he fears boring them) the hardest thing to communicate is the weight of having to fill all those hours locked in that room in Gaza. This is how he tells people they could understand it. Place one white, plastic chair in the middle of the room. Then sit there for three hours. Then for another six hours. Then for three hours more. Then remember that there are still four hours to go before you can allow yourself to fall asleep to relieve the boredom. If you were really to try this there would still be one major difference between your experience of time and his – you'd know you could give up at any point you choose. Alan didn't have that option; he knew that tomorrow would be exactly the same. As would the next week. And the week after that. Maybe for years.

It was clear to Alan that in order to deal with his 18 waking hours each day, it was essential to harness the fact that we construct our own perception of time:

'After about eleven days there was that shock-of-capture period, when you think this is absurd, you can't go on. This isn't going to happen. And then there are moments when you think, Christ, I'm the Brian Keenan of Gaza. I remember that eleventh night. I'd just had a wash. I sat on the chair and thought that I'd arrived on firmer psychological ground at that point. I thought, this is the long one. I could be here for three years. I'm generally quite pessimistic. If I can expect the worst scenario in my head then everything will seem easier. So I decided to be ready to take three years in captivity and that anything less than three years would be a huge bonus.'

Alan cleverly adopted a strategy of considering life through dual time-frames. While assuming he would be captive for three years, on a day-to-day basis he told himself the experience could end at any moment. 'Every single evening when the call to prayer came I used to say to myself, almost out loud, this wasn't your day, but maybe tomorrow will be.'

On 4 July 2007, after almost four months, the Army of Islam group holding Alan Johnston hostage handed him over to officials from Hamas. His ordeal had ended and soon he would be free to return home. It was at the start of his journey back to Scotland that he noticed that something had changed in the way he experienced time.

'On the flight home from Israel somebody got a small dog through security. This woman sat there with a Chihuahua. When the staff realised the dog was there, we had an hour's delay while they removed it. Everybody was so annoyed that it was slightly surreal. I couldn't work out why they couldn't bear waiting for just for one hour. Yet within six weeks of getting back to London I remember waiting at a bus stop and swearing because there weren't any buses coming along. Already the old impatience had come back. I'd hoped I would avoid that. After everything my parents had gone through, and all that hassle for the BBC, I wanted to take something useful from it. It's like walking on air when they let you go. Everything seems so fantastically good. If only you could hold on to just one per cent of that appreciation of freedom, but you very quickly see it dwindling away.'

When I interviewed Alan Johnston he was waiting to hear whether heavy snow would prevent him from reaching Scotland for Christmas. It wasn't looking hopeful, but his reaction suggested that despite what he said, he had taken something lasting from his experience. 'If I don't go to Scotland for Christmas it's not the end of the world,' he told me. 'When I was in captivity I'd have given anything to be stuck in London unable to get home for Christmas, or sitting on a plane waiting for them to offload a Chihuahua.'

Luckily most people won't encounter a situation as appalling as Alan's, but his experience does demonstrate the flexibility of our experience of time. If he can make time in captivity go faster, then it must be possible for the

rest of us to speed up our experience on something inconsequential like a long-haul flight. To do it, you need to make
every attempt to avoid all the factors which are known to
decelerate time, which is of course what most people try
to do. They try to get comfortable and then they partake
in that hour-engulfing activity so despised by time
researchers – watching TV. It works because anything that
absorbs you or distracts you from the passage of time itself
will make it speed up. So you should avoiding checking
your watch too.

But what if you find yourself in a situation without such
distractions? You're on a broken-down train with nothing
to read, no signal on your phone and no one to talk to. In
this situation you need to do the opposite: distracting yourself from your surroundings is not going to work, so try
focusing on them instead. This is where mindfulness comes
into play again. Taking the senses one at a time, observe
everything in the carriage. Notice all the different textures
– the smooth, shiny poles, the slightly furry seats, the metal
ridges on the floor. Then there are the smells, the sounds,
the sights. If you can look on this as an opportunity to
practise 10 minutes of uninterrupted mindfulness then
you'll feel less irritated. The more absorbed you become,
the faster time will go.

PROBLEM 3: TOO MUCH TO DO, TOO LITTLE TIME

The invention of the car hasn't saved us hours of travelling; instead we travel further. Social-networking sites
haven't saved us time seeing people; instead we stay in

touch with more people and communicate with them more often. When I learnt to edit radio programmes I used white sticky tape and a razor blade. We would sit and edit with long strings of black tape hanging round our necks. We would sometimes cut our fingers by mistake and became accustomed to sorting patiently through spaghetti-like tangles of tape on the floor to find the perfect piece when yet again we had dropped it. There was no doubt that it was more time-consuming, but now that today's digital editing means we can edit much faster it has also allowed us to become fussier, removing every 'um' and 'er' and experimenting more with the order of a piece. The result is that it takes us just as long.

Despite all the new technology we have, many of us still feel there aren't enough hours in the day, and that if only there were, life would be easier. There is some evidence that the number of hours when we feel forced to rush has more influence than age does over the perception that time is moving fast. In an internet study of more than 1,500 people in the Netherlands, the psychologist William Friedman found that those who felt they spent a lot of their time racing to do everything they needed to, also believed time went very fast.[126] The consciousness of not having enough hours draws our attention to time slipping away, making it feel faster.

The world of time management wants to come to the rescue, with its promise of improved productivity and the prospect of personal transformation by the saving of so many hours. Along with increasing our efficiency at work, we will suddenly find we have the time to learn new languages, get fit, bake our own bread in the morning, run a small start-up from home in the evening and charm friends with the hand-made gifts we make for them at weekends. The only problem is that however clever some of the techniques of time management might appear – software that analyses your computer-use second by second, digital alarms which marshal the minutes, advice on triaging your tasks, not triaging your tasks, setting goals, assessing urgency versus importance or even timing tasks to fit in with your 'natural rhythms' – there's very little researched evidence to suggest that adopting these techniques makes any difference at all.

Some swear by starting the working day doing a single task for an hour before opening their emails. This gives them the satisfaction of completing a substantive task early in the day, ahead of an activity that invariably results in a list of new tasks. Others find lists help them to prioritise, and serve as a memory aid, but of course this only saves time if you don't spend more time colour-coding the list than working. Some keep a 'done' list, adding tasks after they've been completed, giving them the pleasure in seeing how much they've achieved in a day. There are tips to deal with the screenfuls of emails waiting for you after a holiday, such as starting with the most recent instead of the earliest in the hope that some problems have been resolved by the time you reach the newer messages, or the riskier strategy of deleting the whole lot on the basis that if anything is really vital either someone else will tell you about it or the sender will email you again.

Any of these strategies *might* work for you, but evidence that they work for everyone or even most people is hard to find. Many people who use their time efficiently don't employ any specific time-management techniques. Yet the quantity of advice available shows that there is a demand for help and that many people do have a desire to fit more activities into fewer hours.

This suggests to me that perhaps we need to address something different – the perception that we have no time. Most people in employment claim to be time-poor, but what if the deficit lies not in hours but in their estimation of free time? In the same way that sleep diaries reveal that people who believe they have insomnia in reality sleep for a bit longer than they realise, activity diaries demonstrate that most people underestimate the amount of free time they have – and underestimate it considerably. In one study people guessed that they had 20 hours spare each week. Their diaries revealed that they had 40. To have 40 spare hours a week would suggest you could fit in a second full-time job, but the problem of course is that not all hours in the day are created equal. Two free hours when you are tired late at night are unlikely to be as productive as two hours during the day.

Even if we concede that we have more spare time than we think, this doesn't avoid the fact that sometimes we have deadlines to meet which seem impossible. What can research tell us about the most efficient way of using the hours when we're up against it? Multi-tasking is a theme that comes up a lot here. Is it quicker to do things all at once or one at a time? If I look at my computer screen right now there are four Word documents open including this one, three PDFs of journal papers, three email accounts, one social-networking site and four other websites. This is partly because I refer to several sources while I work, but also because I can't resist social contact, even when I know it distracts me.

It seems I'm not alone, and that this trend is increasing. The younger a person is, the more likely they are to employ two forms of media simultaneously. In the early evening a third of people are using two at once, for example, chatting on the phone while surfing the internet or texting while watching TV.[127] In theory this could save time, like the British cabinet minister who confessed that the job was so pressured that she saved vital minutes by cleaning her teeth while she was on the loo. The alternative view is the mono-chronic assumption – that it is always better to complete one task before beginning the next. In research conducted over several decades, Allen Bluedorn has found that, unsur-prisingly, it's a matter of personal preference. Some people favour monochronicity and feel happier completing one task before they start the next. Others are polychronic and do appear to perform better when they are doing lots of things at once, and can excel in jobs that require them to do just that.[128] Running a busy café would be a good example – though this doesn't mean they necessarily get the jobs done faster. In a café there's no option but to jump from task to task. However, if your job does give you the choice, then it's as well to be aware of what's known as attention residue. When you switch from one task to another, it can be demonstrated experimentally that a bit of your mind is still focused on the previous task. Each time you switch back again you have to remind yourself about what it was you were doing, while dealing simultaneously with the slight distraction from the first.[129] Although this can increase your cognitive load, many people still prefer to work this way and there is no problem with that. It only creates difficulties if you feel unable to focus on any single task. Then some people find that setting an egg timer for between 15 and 20 minutes, and deciding to focus on one job until it rings, can help them to concentrate. It might work for you, and it does for me, but when you do it repeatedly it can be a very intense way of working and, again, it is hard to find much more than anecdotal evidence to support the idea that working this way is beneficial. The research really hasn't offered up a single time-management prescription that will work for everybody.

Research on future thinking has revealed that deadlines do strange things to our experience of time. In the same way that a building looks further away when you're carrying a heavy suitcase, the psychologist Gabriela Jiga-Boy found that an event seems more distant the more you need to do before you reach it.[130] But only when there's no deadline. When there's a firm deadline everything changes; it makes the event feel closer. So if you are house-hunting, the actual day of the move may seem far away because you know there's so much to do beforehand, but if you have a deadline like trying to move house before the birth of your baby, the date will feel much closer.

Deadlines do strange things to the mind. They can even reduce the attention-residue problem, which can occur when moving from one task to another. When you complete a task up against a deadline, you are forced to narrow your options and to make decisions that are cognitively less complex. This in turn decreases the hangover from that first task, allowing you to put it behind you and get on with the next job. So an approaching deadline not only concentrates the mind, but allows it to clear more easily after it's passed, leaving us to worry about the next deadline instead.

If, after experimenting with the use of deadlines and monochronic versus polychronic work, you still find yourself with too much to do and too little time to do it, then you have a choice. Either slim down your commitments or accept that you are busy and are likely to be so for a long time to come. We tend to kid ourselves that if only we can get through this week or the next month, things will be calmer in the future. This might be true if you have a big one-off project, but experience probably tells you that it's not. Forever yearning for that calm future where everything is perfectly organised sets you up for disappointment. You probably won't reach this imaginary period of order and relaxation. Unforeseen events will continue to happen to members of your family, computers will go wrong and something in your home will always need mending. And if you do find yourself with that longed-for, undisturbed stretch of time in which to relax, it might not even make you happier. Research on British ex-pats who had moved to south-west France found that once people had finished working on their houses they became less happy because they had nothing to do. They were now living in the *gîtes* or chateaux they had dreamed of and worked so hard to restore, but time hung heavy. Maybe there's only so much pleasure to be had relaxing with a glass of rosé on the perfect terrace. As the researcher put it, if you're moving abroad for that place in the sun, the lesson is: never finish your house. So, unless you feel you can't cope, maybe it is best to come to some accommodation with time, to accept that your timetable is full and will continue to be so. And think of the advantages instead – a full schedule will create plenty of memories on which you can look back, reducing the impression that time is rushing by.

Or maybe you decide that you do need more spare time, in which case I like Philip Zimbardo's recommendation that you start viewing time as a gift and choose whom to give it to. If time feels scarce, choose to give it to two sets of people – those who gain the most from spending time with you and those whom you most want to see. Refuse some invitations. I have to confess to liking the episode of *Friends* where, faced with someone needing help moving house, Phoebe simply said, 'I'd love to help, but I'm afraid I don't want to.' Not very generous perhaps, but honest, at least.

If you do make a decision to clear more time in your diary, there is one other factor to take into account. Mark Williams, a clinical psychologist at Oxford University who researches the psychological benefits of mindfulness, has noticed that when people feel stressed and overwhelmed by their busy lives, they often choose to give up the one activity that most benefits their well-being. It is easy to see why. They can't choose to give up their families or their jobs, but they can stop singing in a choir, taking exercise or going to evening art classes. These are added extras that might seem hard to justify time-wise, but in fact have been demonstrated to reduce stress and increase well-being.

One final word on this topic. With all today's talk of work-life balance and 24-hour schedules, it is worth remembering that the pressures of time are not confined to modern life. In 1887 Nietzsche described a feeling that seems familiar now: 'One thinks with a watch in one's hand even as one eats one's midday meal while reading the latest news of the stock market.' A compilation of five different measures of time taken over the past 50 years indicate that the average American man has six to nine hours *more* free time every week than he had five decades ago. The American Time Use Survey from 2010 revealed that men have 5 hours and 48 minutes of spare leisure time each day, while women have a little less (funny, that), with 5 hours and 6 minutes of leisure time each day. The survey also reveals that if people find their leisure time increases – perhaps they have found a way of working more efficiently or have reduced their commitments – then there is one chief way in which they use that longed-for extra free time. They watch more TV. So if the political scientist Robert Putnam was correct in his assertion that each extra hour people spend watching TV is associated with lower levels of social trust and group membership, then could more free time counter-intuitively result in lower levels of social engagement?[131]

PROBLEM 4: FAILING TO PLAN AHEAD

Sometimes, however organised you are, however much you slim down your commitments, you cannot meet your deadline, even when you chose the date yourself. Here the

Planning Fallacy is at work and it means the deadline you set was never realistic. The Planning Fallacy is the tendency to believe that a job will take less time than it eventually does. If you spot that this is something you are prone to, you can avoid it. Combining the findings of all the research that's been conducted in this area, here's how to devise a realistic time-frame for a task: list every concrete step and estimate how long each will take; think of past events and look for similarities and differences and add on some more time if they took you longer in the past; add on some more time for anything you can recall that disrupted you last time and finally include some extra hours (or days, depending on the total duration of the project) for the unexpected to happen. Then look at your diary and calculate precisely how many hours you have available to devote to the project, bearing in mind you will have no more free time in the future than you do next week. Only after all that can you come up with a realistic deadline. The hardest part is resisting the temptation to give in to the optimistic idea that you will have more spare time in the future. Even the idea of cooking supper for friends one night next week will seem easier time-wise than the idea of doing it tonight. As a final check, since research has shown that *other* people make more accurate judgements about *our* time, describe the task to a friend and ask them to guess how long it will take you. The more skilled at deadline-setting we become, the less we are forced to rush because we stop over-committing and then fearing we'll let other people down when we can't keep up.

When we plan ahead, the major finding from the work

on future thinking is the tendency to ignore the inessential features of an impending event. Again there's a simple way round this. If you feel you are someone who takes on too much (and this might not apply to you – I'm not saying everybody should turn down every request), then before you commit to an event later in the year, imagine it is happening next week. If it seems out of the question that you could fit it in, then ask yourself what steps you would need to take to be free to do it in six months' time, remembering once again that you are unlikely to have more free time. By imagining it is next week you are more likely to consider the practical feasibility of the whole event, instead of focusing on the main part of it.

Consciously deciding to plan ahead in detail in your mind can even bring feelings of calm. The happiest people tend to imagine a greater number of separate steps in their future plans, even in something as trivial as a trip to the supermarket. They seem able to picture every detail of the trip in their minds.

This book has focused on the way we – as individuals – experience time, and how that can change our lives. But the same principles can be applied to the bigger picture. There are many examples where the ideas discussed in this book could be used to shape policy decisions. To prevent expensive over-runs on capital projects, part of the procurement process should involve a third party who analyses the factors that held up previous projects, assesses their similarity to the current project and then estimates the completion date for the project. They should be entirely neutral and separate from the bidding process. This would minimise the tendency of companies to give over-optimistic predictions of completion dates. It might feel like a waste of money to employ a consultant to do this, but when the evidence reveals our inability to make these judgements for ourselves, it could save millions or even billions.

If, as an individual, you have a tendency to plan, but then forget to actually do the things you meant to, then research on future thinking suggests a simple way of remembering that really does work. Imagine yourself carrying out those tasks and the steps you will need to follow in order to do them. So if you need to post a letter on the way to work and to buy some washing-up liquid on the way home, rather than trying to remember that you have two tasks to do, picture yourself actually doing them. Imagine which post box you will visit and when, and picture yourself putting that exact letter into the box. This only need take a second while you put the letter in your bag. Then decide where you will buy the washing-up liquid, visualise yourself finding the correct aisle, choosing a brand, picking it up and queuing to pay. This is far more effective than repeating the words to yourself in your head. It won't work every single time, but you'll be surprised at how often it does.

This technique can also help you to keep to any resolutions you make. If you not only make a plan, but imagine yourself carrying it out, you are far more likely to stick to it. Researchers have found it even works for encouraging people to eat more fruit. Given the goal of eating extra apples and bananas over the subsequent seven days, the students who were told to visualise when and where they would buy the fruit, and how they would prepare it, increased their fruit consumption by twice as much as those who simply set out to eat more fruit. The crucial element in successfully using imagery is to imagine the process and not just the result. Imagining holding the trophy high won't help you win Wimbledon, but envisioning how you will approach playing perfect shots could.

It is clear from the experiences both of Alan Johnston in captivity in Gaza and Victor Frankl in a series of Nazi concentration camps, that imagery can bring solace in the most extreme situations. I'm struck by the fact that they both made active decisions to retain mastery of the one element of their lives over which their captors did not have complete control – the state of their minds. They were both determined to use their own thinking as a mechanism for coping. In 1945 Victor Frankl wrote his memoir of his time in the camps, *Man's Search for Meaning,* in just nine days. It has sold more than 9 million copies worldwide, a figure which he found somewhat perplexing. Up until the last moment he had been intending to publish it anonymously and was puzzled that of the dozens of books he wrote, this should be the one which made him famous. Frankl's preoccupation with controlling his own mental state gave rise to a type of talking therapy called logotherapy. Frankl reasoned that if a person experiencing the terror of the holocaust could find a way of controlling their mind then this must be possible in everyday life as well.[132] Frankl wrote, 'Between stimulus and response there is a space. In that space is our power to choose our response. In our response lies our growth and our freedom.'

While he was in the concentration camps, Frankl deliberately escaped the horrors of the present by throwing his
mind forwards into the future. There was one type of
suffering he found worse than any other. Although he was
constantly cold, aware that he was slowly starving and lived
with the continual fear of death, it was time that terrified
him the most. He found not knowing how long he would
be there to be intolerable. This lack of a time-frame for the
future was 'the most depressing influence of all'. The
moment people arrived they said they knew they had no
future. Some coped by closing their eyes and perpetually
living in the past, but Frankl was convinced that the only
way to survive was to plan – to somehow find goals, however
small, which could give him a semblance of a future. In
one of his lowest moments when he was marching in the
bitter cold with sores on his feet, he forced himself to
imagine he was in a warm lecture theatre giving a talk on
the psychology of the concentration camp. This imagined
future enabled him to complete the walk. He was manipu-
lating his own mind time in order to survive.

PROBLEM 5: A POOR MEMORY FOR THE PAST

It is inevitable that memories fade over time, and for people
whose memories are as traumatic as Frankl's this can sometimes be welcome. But it is the flexible nature of memory
that gives us such a strong power of imagination, so when
we do forget positive memories too, we should not be too
hard on ourselves. However, there are occasions when we
wish we could remember more, and the research on the

psychology of time perception does give us ways of improving both our memory for events and our ability to guess the date of those events correctly. The issue of dates is easier to remedy. By looking at the mistakes people made when they were tested on the contents of their daily activity diaries, I have devised a three-part system for estimating the date of an event more accurately – useful for everything from the trivial, such as finding out whether a broken kettle might be under guarantee or guessing when you last visited a friend, to the more serious, such as retrieving an old piece of work or giving reliable testimony in court.

First make a rough guess as to how many weeks, months or years ago the event took place. Next try to guess the actual date. This leads to more accuracy than guessing periods of time. Then to make your final estimate, add or subtract some days, months or years according to the following rules: for a personal event which happened more than two months ago, add some time on. So if it was six months ago that you think you went to France, it was probably seven. Eight years ago was more likely nine. But if it happened less than two months ago then the likelihood is that it happened more *recently* than you think – this is reverse telescoping. So if you guess it was roughly ten days ago, take a day off and go for nine days instead.

For public news events the rules are slightly different. Remember that key figure of 1,000 days or three years. If you think an event happened three years ago then you might well be correct, so stick with that guess. If you know it was less than three years ago, make it a month or two more recent than your first thought. If it was more than three years ago, perhaps even decades ago, make it a year or two or three earlier to compensate for the telescoping of time.

You can also look for time-tags; the personal events that tether our memories of the news. We know that people who are best at dating events are those who are able to link them to what was happening in their own lives. So to work out when Princess Diana died, consider where you were when you heard, or when the funeral took place. Where were you living? Where were you working? Then focus on any details which might give you other clues. What was the weather like when people were leaving flowers in Kensington Gardens? Was it light in the evenings? Could it have been in summer or winter? You would do some of this instinctively, but the more you deliberately pay attention to the details, the more likely you are to be correct. As you might expect, it's the most significant events in your personal life that help you to tether a memory – so did it happen before or after you had a child, or before or after you moved house?

Then there's the more general problem of forgetting events entirely. The more you discuss an autobiographical memory, the more likely you are to recall it, but that doesn't mean that you will remember it accurately. Each time you tell the story you can reinforce earlier mistakes, but there are methods of improving recall. Think back to the research John Groeger did where he asked more than a thousand drivers to describe past car crashes, however minor. People remembered more accidents when they were told to work from the past to present, rather than the other way around.

And they recalled more if they started from a specific date in time, rather than racking their memories for accidents from within a fixed duration such as the last year. This is a strategy anyone can use. If a job application form asks you for specific occasions where you faced up to a particular type of challenge at work, recalling them is a difficult cognitive task because you won't have categorised these memories in your mind in that way. But again the research findings can be put into practice. Don't think of the most recent challenge working backwards through time. Start with the first job you had that's relevant to your application, think of those first few months and whether there were any challenges there, then move on to the period where you were more confident and might have begun suggesting challenging projects at work. Do the same for each job up to the present day and the likelihood is that you will come up with far more examples than if you worked backwards. If you get as far as an interview for the job and further questions of this nature come up, you can do the same, but faster of course.

A final tip is to start with a longer time-frame than the one you are aiming for. So if a visa application asks you how many times you have been abroad in the past three years, think back five years and then narrow it down to three. This avoids the problem brought on by telescoping, where events are included from beyond the time period in question.

There is also a wider issue here for policymakers. The accuracy of the public surveys used to inform government could be dramatically improved by taking research on telescoping into account. If, for example, people are asked to look back and say how many times they have used their local swimming pool in the past three years, then it is important to know how the psychology of time might alter the accuracy of their recall. The survey should start with a longer time-frame and then narrow it down, and it should give people fixed dates to work from, starting with the earliest and working forward to the present day. This will give more accurate responses and minimise the influence of telescoping on the reliability of the results – essential for a government trying to get an accurate assessment of the use of different public services.

PROBLEM 6: WORRYING TOO MUCH ABOUT THE FUTURE

Daydreams can be enjoyable, and drifting into the future could even be the brain's default position when it's not occupied. But when daydreams turn into obsessive worrying they are far from satisfying, and excessive rumination can have serious consequences. When we think about the future, we combine old memories to create plausible future scenarios, but sometimes the plausibility seems to be missing. We begin catastrophising and thinking only of the worst. There are various approaches to managing excessive worry. A classic strategy from cognitive behavioural therapy involves *deliberately* imagining the worst-case scenario concerning your worry, followed by the best possible scenario. The real outcome probably lies somewhere in between. To take an example, if you are dreading telling your boss that you've made a serious mistake at work, your worst-case scenario is that they scream at you in front of everyone and fire you on the spot. The best is that they say it really doesn't matter and that they're glad you stopped by because they had been meaning to tell you about the pay rise they wanted to give you. The most plausible scenario probably lies between these two, but imagining them both can help you envisage the most realistic outcome, allowing you to worry a little less.

Ad Kerkhof is a Dutch clinical psychologist who has worked in the field of suicide prevention for 30 years. He has observed that before attempting suicide people often experience a period of extreme rumination about the future. They sometimes reported that these obsessive thoughts had become so overwhelming that they felt death was the only way to escape. Kerkhof has developed techniques which help suicidal people to reduce this rumination and is now applying the same methods to people who worry on a more everyday basis. He has found that people worry about one topic more than any other – the future, often believing that the more hours they spend contemplating it, the more likely they are to find a solution to their problems. But this isn't the case. His techniques come from cognitive behavioural therapy and may sound remarkably straightforward, but they are all backed up by trials.[133] I like the fact that he does not make grand claims for his methods. He openly told me that anyone who thinks these techniques will banish all worrying will be disappointed, but that if a person hopes to cut down the time spent worrying, then this is achievable.

If you find yourself awake in the middle of night worrying, with thoughts whirling round repeatedly in your head, he has several strategies you can try. This is where imagery comes in useful again. Imagine there's a box under your bed. This is your worry box. As soon as you spot thoughts that are worries, imagine taking those individual worries, putting them into the box and closing the lid. They are then to remain in the box under the bed until you decide to get them out again. If the worries recur, remind yourself that they are in the box and won't be attended to until later on. An alternative is to choose a colour and then picture a cloud of that colour. Put your worries into the cloud and let it swirl backwards and forwards above your head. Then watch it slowly float up and away, taking the worrying thoughts with it.

This might begin to sound like psycho-babble, but there is strong empirical evidence from Kerkhof that for some people it does work. Not everyone finds that imagery suits them, but Kerkhof has another technique that he finds to be effective for most people. This is to set aside a time for worrying. Your worries relate to real and practical problems in your life, so you cannot rid yourself of them altogether, but you can learn to control when you think about them. Fyodor Dostoyevsky famously commanded his brother *not* to think of a white bear, and we know from the experiment on thought suppression which followed that, given that instruction, you can think of nothing but a white bear. I knew the feeling when I was a guest sometimes on the old TV chat show *Richard & Judy*. Because the programme was sponsored by

Schwartz spices, strict sponsorship regulations meant that any mention of this particular brand of spices during the programme itself was banned. So every guest was given a piece of paper to sign beforehand promising not to refer to you-know-what-kind of spices live on air, with the result that, hard as you tried not to, they were all you could think about. Likewise, telling people *not* to think of their worries isn't going to work. Instead Kerkhof recommends the opposite. Set aside 15 minutes in the morning and 15 minutes in the evening to do nothing but worry about the future. Sit at a table, make a list of all your problems and then think about them. But as soon as the time is up you must stop worrying, and whenever those worries come back into your head remind yourself that you can't contemplate them again until your next worry time. You have given yourself permission to postpone your worrying until the time of your choice. Remarkably, it can work. It puts you in control.

PROBLEM 7: TRYING TO LIVE IN THE PRESENT

Back in 1890 William James mused on how to live in the present. 'Let any one try, I will not say to arrest, but to notice or attend to, the present moment of time. One of the most baffling experiences occurs. Where is it, this present? It has melted in our grasp, fled ere we could touch it, gone in the instant of becoming.'[134]

We might talk of our desire to live in the present, to stop our wandering minds from constantly leaping ahead or trailing back into the past. But do we really want to be stuck in the present like H.M., the man whose hippocampus was sucked out through a silver straw? It is true that increasing our attention to what is around us through techniques like mindfulness or meditation can bring benefits in terms of well-being, and this is something I will discuss in more detail later in this chapter, but there is a limit to how much we should desire to live in the present. On hearing the story of H.M., most people feel sorry for him, living in a nursing home, with no new memories, unaware that he was the most famous patient in neuroscience. He was condemned to live in the present, yet here are the rest of us, feeling we ought to be doing it more. Small babies are very good at it. They have the ability to learn, yet don't have long-term memories and have no idea what might happen the following day, let alone the following month. The consequence of this is a total lack of control over their lives or the way they spend their time. They can't plan; they can't anticipate the following week; they can't even look back fondly on earlier experiences.

As we grow up and our brains develop, time-travelling into the future appears to become the default position when the mind is wandering. Instead of yearning to live more in the present, maybe we shouldn't fight the lure of the future. Research has found that anticipation is associated with stronger emotions than remembering the past, so if we want to improve our well-being maybe we should pay less attention to the pleasures of nostalgia and more to anticipating positive events in the future.

But what if you are convinced that you already spend

too much time pondering the past and the future? James is right that the present moment constantly gets lost, but there are ways of stopping your mind wandering forwards and backwards in time. 'Flow' is the name for a special state of mind identified by the psychologist Mihaly Csikszentmihalyi. It means you become so absorbed in the task at hand that you think about nothing else and soon have no idea how long you might have been doing it. This is different from simply being very busy. Your mind is completely focused on the activity at hand; it's not drifting back into the past nor forwards into the future. If you can find the activity that does this for you, you can slow the passage of time without feeling bored. It might come from playing music, running or gardening. You might already know what does it for you. For me it's painting. When I travel I take a small sketchbook and paint when I get the chance. Very quickly I become completely absorbed in the picture and the view I'm painting. I probably feel calmer than at any other time. You lose all self-consciousness and become completely immersed in the process of the activity itself, rather than the outcome. Csikszentmihalyi identified certain essential features for a situation to induce flow. It needs to be neither so easy that you don't need to concentrate, nor so difficult that you worry about the results. The activity should have clear goals, over which you feel you have some personal control.[135]

Philip Zimbardo conducted an experiment that coincidentally focused on painting. A group of people were asked to paint a picture of a basket of flowers. Half were told the painting would be judged afterwards by graduates from the art department. The other half were to concentrate on the creative process itself and not to worry about the final product. When the art graduates judged the paintings they rated the second group's more highly, despite the fact that they weren't focusing on the results.[136] This suggests that this state of mind might not only keep you in the present but has benefits for creativity too.

However, the concept of flow presents us with a paradox. When you are experiencing flow, time does seem just to disappear. Hours can pass and you don't notice them, so if time going too fast is your main issue then you might not want to spend your days experiencing flow. On the other hand there is evidence to suggest that flow increases well-being, so perhaps once again our perception of the speed of time is less of a problem than we think.

Holidays are occasions where we often make a conscious effort to try to live in the moment. We have worked for months to afford a break from the routines of home and office, and we want to make the most of it. But how often have you found yourself wandering around somewhere on holiday imagining what a great place it would be to visit on holiday and how you might spend your days? Then you remember that you already are on holiday and should be enjoying it right now. On a break in Spain I decided to investigate how long I could truly appreciate the here and now – putting thoughts of the past and concerns about the future to one side. The answer is: not very long, even though I deliberately chose to immerse myself in a moment of profound serenity and beauty.

Here is my experience: I am staying in a lovely bed-and-breakfast outside a village in the hills of Southern Andalucía. It is a simple, though stylishly restored old farmhouse, with a small infinity pool that possesses one of the most spectacular vistas in Spain. In one direction I look up to the little *pueblo blanco* clambering up and over the escarpment. Behind the white-washed houses is a craggy sierra silhouetted in shades of purple against the aquamarine of the sky. I look out on a dry, deep green valley of olive trees and vines, behind which is a range of rolling hills, studded with white wind turbines, turning in the soft breezes like Quixote's windmills. Beyond the hills is the Mediterranean with the oblong dots of impossibly slow-moving tankers and then the Rock of Gibraltar standing sentinel over the narrow straits. Far in the distance is the shimmering haze of the mountains of northern Morocco – with their promise of a whole other continent.

Yes, I am treading water in this little pool in a Spanish garden while staring out over Africa! If I look down, this stunning view is reflected on the surface of the water, and if look to the side there are well-watered lawns, olive trees, lime trees and psychotropic orange datura flowers, buzzing with insects and dancing with birds.

This should be the perfect place to luxuriate in the sensory overload of now. Why would I want to look at or think about anything else other than this moment in all its gorgeous fullness? It isn't as if I have any particular worries or concerns at this moment. Life is as good as it can be. All I have to do is to make the most of it.

So, leaving the swimming pool, I lie back on a sun bed

and try to take it in, to appreciate it fully, to avoid any distractions or lapses into memory or speculation. And yet after a few minutes I am reaching for a book. I choose a guidebook of Spain, so at least I am not being transported too far away, as I might be by a novel, perhaps. I find myself reading about this very place – the descriptions of the guidebook writer helping to validate my immediate experience. But why do I need someone *else* to tell me what I am feeling right now? What can his descriptions add to my own senses? And yet, somehow, it helps to confirm me in my happiness that I am here, now. Then, without being able to stop myself, I start looking at all the other lovely places there are nearby. That lake, this gorge, this village, that hotel – and before I know it I am planning where to go later. Already this moment on the sun bed is turning into the past, is being ecstatically memorialised, is becoming a gushing anecdote for home, accompanied by photos – 'We stayed in this lovely place. It was amazing!' – and I am projecting myself into future plans. I am thinking of lunch and whether we should walk up to the village afterwards. Then there's dinner. Someone's told us the food at this place is very good. Or we could eat at the hotel up the road with the famous restaurant. There is a pleasant anticipation in this, but all sense of the now, with all its possibilities, is lost.

When the philosopher Alain de Botton talks about travel, he says that however rich and enjoyable the experience is, it is somehow ruined by the fact that we bring ourselves along for the ride. I would argue that it is not just ourselves that we bring along, but our past and our future too. And however much we are enjoying the now, we can't stop thinking about what we have left and where we are going next.

Even if you can find an activity that provides you with flow, there might not be the time or opportunity to do it very often, so an alternative is to practice mindfulness. Earlier I discussed using it to fill the time on a broken-down train. It is a form of mental training where you learn to stop your mind hijacking your attention and sending you into the past and future when you don't want it to. Instead you learn to focus your attention. The advantage of mindfulness is that it is a method of bringing your mind into the present wherever you are. It makes time pass more slowly, but also more pleasantly. Although mindfulness has been practised for centuries as part of Eastern spiritual traditions, recently there's been a growing interest from professionals within the fields of clinical psychology and neuroscience. It has been found to be particularly useful for controlling the wandering mind when people are feeling depressed or anxious. In trials of mindfulness-based cognitive therapy Mark Williams, the Oxford clinical psychologist who is a leading figure in this field, found that an eight-week course consisting of a two-hour session each week halved the recurrence rate in people who had previously experienced more than three episodes of depression.[137] Encouragingly, he found it particularly effective in the people whose depression was proving hard to treat.

However, it is a skill that can be used by anyone. The idea is that you learn the ability to focus your mind whenever you choose to. You will want to select your moment carefully. If you're taking a gondola trip through Venice, the last thing you want to do is focus on your breathing and how your body is feeling. But at other moments it can make you feel much calmer and stop thoughts from the past and future from crashing in on what you are doing. Trying it for just 20 or 30 seconds a few times a day can make a difference and crucially there's a growing evidence-base for the practice. As well as the documented improvements in people with depression, measurable changes can be seen in the insula in the brain, which senses the state of the body and your emotional state. Differences are also observable when it comes to the parts of the brain controlling attention.

Having read the impressive research, I decided it was time to take a lesson myself. My teacher was lecturer and therapist Patrizia Collard, and since I feared my commitment might waver if I had to invest extra time in practising, her task was to train me to do it on my morning walk to the station. I walk up the same, rather dull street of boxy houses where the only thing to occupy my mind is the bafflingly high prices they sell for. According to Patrizia this slightly dreary road is the perfect place for mindfulness. First she told me to ground myself, to stand still and feel my connection with the ground. I had the choice between seeing myself as an oak tree, a mountain or a sumo wrestler. Or you can think of anything else that is broad, strong and stable. I became aware that my feet were connected to the ground. Then, breathing slowly and deeply, I noticed how I began to feel calmer. We focused on the in breaths and the out breaths and then started walking. She suggested breathing in with the first step and out with the third, but this meant walking slowly, which would make me late. So it's best to experiment until you find your own speed and work out how footsteps correspond to the breaths.

Once the breathing was co-ordinating neatly with my footsteps, I was to take in the environment, paying attention to the objects around me before selecting something pleasant to look at. I chose a tree, and noticed the surprising number of different shades of green and yellow there were amongst its leaves. Meanwhile the breathing rhythm is supposed to continue, but that takes some practise. Then you ask yourself how your body is feeling right now. Are there aches and pains? Are your shoulders hunched? Yes. Can you relax them? Maybe. Are you frowning? Probably. Smile to relax the face. 'A smile is like champagne in the body,' she told me. Now go to work, she said, and you'll arrive thinking, 'I'm ready for this day.'

I have to confess that I don't remember to do it every day, but when I do, it does make a difference. The idea is to arrive at work feeling calm and focused, instead of making the journey on auto-pilot, while troubling your mind with all the things you need to do that day.

PROBLEM 8: PREDICTING HOW YOU'LL FEEL IN THE FUTURE

When it comes to imagining the future, copious research on the Impact Bias has illustrated how poor we are at predicting our emotions. Luckily the reasons for this are clear, which means there are ways of overcoming it. If an event is happening next year, imagine instead that it is happening next week and ask yourself how you would feel about it then. We have a tendency to focus only on the most extreme features of an event, so if for example you are trying to imagine what life would be like for you if you changed job, remember to consider not only the job itself but all the other features that will affect how happy you feel in that job. Will you have more or less control over your own time? Once the excitement has worn off, which elements of the job will bring you satisfaction? If it involves a pay rise, is there something you will do with that money that will make a difference to your well-being? Who will your colleagues be? How does the atmosphere there compare to that in your current job? And then remember that your life outside work will still affect you. How much of that will have changed too? Will the journey to work be easier or harder? Will you still be living in the same place?

Surprisingly you might not be the best person to assess how much you'll enjoy this new job. If you really want to know how you will feel about a future event, Dan Gilbert, a psychologist who has done a great deal of the research in this area, recommends abandoning your imagination altogether. Simply ask someone who already does that job how they find it. What are the best and worst parts of it for them? He concedes that people rarely like this advice. Why believe the view of one person who might be nothing like you? But his research does show that, a year later, even the assessment of a stranger proves to be more accurate than our own predictions.

It is also governments that need to take note of our difficulties in imagining how we might feel in the future. To persuade us to make adequate pension provision, we could be encouraged to imagine in detail how we would manage life on a small pension not in 30 years' time, but next week. Such an approach would reduce the effect of the tendency to believe that in the future we will have somehow have more money and that we need not worry. This could succeed in persuading – or scaring – more of us into saving more for our old age, thereby reducing the impact on the public purse and the social costs of large numbers of elderly people living in poverty.

Policymakers are inevitably concerned about the public reaction to their proposals, but the psychology of time perception suggests that, should they choose to, they could be bolder. The public response to a proposal will be strongest when it is first announced. Cognitive processes will cause people to focus on the main features and the initial impact a policy will have on them. Research on the Impact Bias tells us that in the future people are likely to feel less strongly. When the ban on smoking in public places was first announced, many smokers were unhappy about it, and used public forums to denounce the policy. But after it was introduced not only did it reduce hospital admissions for heart attacks, as had been hoped, but many smokers found they didn't mind as much as they expected to. In Ireland, for example, the support for the ban among smokers went from only a third up to almost two-thirds a year after the ban. Politicians and policymakers can learn from the psychology of time that media storms and public outrage can quickly blow over. If they hold their nerve and trust to sound evidence-based decision making, they can be ambitious.

IN CONCLUSION

'Life can only be understood backwards, but it must be lived forwards.'   Soren Kierkegaard

'Time rushes towards us with its hospital tray of infinitely varied narcotics, even while it is preparing us for its inevitably fatal operation.'   Tennessee Williams

'I never think of the future. It comes soon enough.'   Albert Einstein

'The time which we have at our disposal every day is elastic; the passions that we feel expand it, those that we inspire contract it; and habit fills up what remains.'   Marcel Proust

A great philosopher, a great playwright, a great physicist and a great novelist, each fascinated by the peculiar nature of the experience of time. There is no doubt that time can appear to warp; that it passes painfully slowly when we feel afraid or rejected, or that it seems to speed up when we are enjoying ourselves or as we get older. The reason for this warping is that our minds actively construct our subjective experience of time through a combination of processes involving memory, attention and emotion. Usually these factors create a smooth, ordered impression of time passing. But only one factor need change for time to appear to skew. If we focus intensely on our present situation or on time itself, whether through boredom or fear, the hours appear to stretch. With the repetition of routine and the creation of fewer new memories, the years appear to flash by. In both the present and the future, it is memory that is fundamental to our experience of time. We forget much of past, we telescope events, we misdate them and we have the impression that – as we get older – time is somehow accelerating. We become accustomed to a certain number of memories fitting a certain time-frame, and when life provides something different our sense of time is knocked off-kilter. This warping of time is amplified by the Holiday Paradox – the fact that we simultaneously hold in mind prospective and retrospective estimations of time passing. When these correspond, time seems to flow normally, but when the equilibrium is disturbed and these two perceptions fail to tally, time feels confusing.

Our experience of time has been revealed to rely on another dimension entirely – space. We don't all picture costumed monarchs or decades in the shape of slinkys when we consider the past, but curiously we do all seem to have a sense of where the past and future lie in relation to the location of our own bodies, a sense that is reflected in the metaphors we use in everyday language. It is this ability to locate time in space that helps us to time-travel mentally backwards and forwards at will, with imagery so strong that it can both provide us with an advantage and put our lives in danger.

We curse the fallibility of memory, but it is precisely this flexibility – as I'd charitably like to call it – that allows us to imagine the future, to imagine anything we choose. It is a time-frame so important for us to contemplate that this could even be the default position of the wandering mind. It brings us opportunities to plan and hypothesise that are unique to human beings. I find it wonderful that we can mentally time-travel in this way, indulging in nostalgia one moment, making plans to change the world the next. But future thinking is a problematic time-frame due to our cognitive tendency to focus on the earliest and the most extreme features of a future event, and to neglect the typical when we try to learn from past experience. The result is that we can make decisions about the future which are very wrong.

When it comes to the brain, the clocks of the mind remain elusive, but despite this we are surprisingly good at estimating the seconds, the minutes and even the hours. No one yet knows exactly how we do it, but it is possible that the brain keeps time by counting its own pulses, pulses that occur when our bodies are carrying out other processes.

The experience of time roots us in our mental reality. Some of us see the future coming towards us. Others feel we are sailing on a river of time that is forever moving on, dragging us with it. But when time warps, we are left feeling confused or worse.

So a greater understanding of how human beings experience and use time can help us to live better lives in more productive societies. These might seem like audacious claims, but there is plenty of time – if we know how to use it to our advantage.

We will never have total control over this extraordinary dimension. Time will warp and confuse and baffle and entertain however much we learn about its capacities. But the more we learn, the more we can shape it to our will and destiny. We can slow it down or speed it up. We can hold on to the past more securely and predict the future more accurately. Mental time-travel is one of the greatest gifts of the mind. It makes us human, and it makes us special.

ACKNOWLEDGEMENTS

THE IDEA FOR this book came out of discussions with my excellent editor at Canongate, Nick Davies, about a slightly different book I'd proposed on the way we hold ideas of the future in the mind. It was Nick who encouraged me to tackle the topic of the perception of time as a whole. With the wealth of research out there, it was always going to be ambitious, but I'm glad Nick had faith in me to be able to pull it together.

Without the academics that have spent years conducting this research, this book wouldn't exist. I'd like to thank the following people whose work has particularly shaped my ideas: Marc Wittmann, Endel Tulving, Dean Buonomano, David Eagleman, Lera Boroditsky, Eleanor Maguire, Jamie Ward, Ad Kerkhof, Katya Rubia, Suzanne Corkin, William Friedman, Daniel Gilbert, Demis Hassabis, Emily Holmes, Daniel Schacter, Donna Rose Addis, Thomas Suddendorf, Karl Szpunar, Philip Zimbardo, Bud Craig, Ernst Pöppel and Virginie van Wassenhove. Writing a book on this topic means you can't fail to notice what a precious commodity time is, and so I'm especially grateful to some of those above who have spent time over the years explaining their work to me personally.

My thanks to Mark Williams and Patrizia Collard for their lessons in mindfulness, to Emmy Goodby for her research for Chapters Two and Four, to Marie McCallum for the index and to Matthew Broome and Dean Buonomano for checking particular sections of the book for me.

Then there are the people who were generous enough to share their own experiences with me – Chuck Berry, Robert B. Sothern, Eleanor and Angela. And I want to make special mention of BBC colleague Alan Johnston, who was not only was prepared to go over his traumatic story once more with me, but had clearly thought very hard about the topic of time before the interview. He could easily have written his own book on his experiences, so it was very generous of him to discuss them with me.

Many listeners to the programme I present on BBC Radio 4, *All in the Mind*, took the time to send in their detailed descriptions of the way they visualise time, and in particular I'd like to thank the following people, who even allowed me to include their descriptions in this book: Clifford Pope, Simon Thomas, David Brock, Katherine Herepath and Chella Quint, as well as others who chose to remain anonymous. And special thanks to Roger Rowland and Lisa Bingley, who took the time to draw pictures of how they see time and gave me permission to publish versions of them.

I've been really impressed with everyone at Canongate, who have efficiency and enthusiasm that writers dream of.

This is a much better book thanks to them. Jenny Lord and Octavia Reeve made tactful yet incisive improvements and have been excellent at spotting mistakes. Thank you to my agents: David Miller for getting the book published, and Will Francis for his detailed suggestions on the text.

Finally, I'm grateful to those family and friends who have put up with my constant complaining that I don't have enough time to write this book, and to my partner Tim who patiently read the lot, improved it a great deal and has endured endless conversations about time.

NOTES

INTRODUCTION

1 McTaggart (1908)
2 Zhong & DeVoe (2010)
3 http://news.bbc.co.uk/1/
hi/5104778.stm

1. THE TIME ILLUSION

4 Described in James
(1890)
5 Husserl (1893)
6 Zerubavel (2003)
7 Bargh et al. (1996)
8 Loftus et al. (1987)
9 Twenge et al. (2003)
10 Shneidman (1973)
11 Broome (2005)
12 Wyllie (2005)
13 Wittman (2009)
14 Cotard (1882)
15 Leafhead & Kopelman
(1999)

16 Baddeley (1966)
17 Hoagland (1933)
18 Halberg et al. (2008)
19 Hunt (2008)

2. MIND CLOCKS

20 Henderson et al. (2006)
21 Koch (2002)
22 Vicario et al. (2010)
23 Craig (2009)
24 Sevinc (2007)
25 Pöppel (2009)
26 Schleidt & Eibesfeldt
(1987)
27 See James (1890) again
28 Lewis & Miall (2009)
29 Zakay & Block (1997)
30 Bar-Haim et al. (2010)
31 Langer et al. (1961)
32 Noulhiane et al. (2007)
33 van Wassenhove (2009)

34 http://www.neurobio.ucla.
   edu/~dbuono/InterThr.
   htm
35 Buonomano et al. (2009)
36 Eagleman & Pariyadath
   (2009)
37 Siffre (1965)
38 Foster & Kreitzman
   (2003)

3. MONDAY IS RED
39 Ward (2008)
40 Mann et al. (2009)
41 See Ward (2008) again
42 Gevers et al. (2003)
43 Cottle (1976)
44 See Cottle (1976) again
45 Boroditsky (2008)
46 Casasanto (2010)
47 Boroditsky (2000)
48 Casasanto & Boroditsky
   (2008)
49 Merritt et al. (2010)
50 Casasanto & Bottini
   (2010)
51 Boroditsky & Ramscar
   (2002)
52 Margulies & Crawford
   (2008)
53 Hauser et al. (2009)
54 Miles et al. (2010)

4. WHY TIME SPEEDS UP AS YOU
GET OLDER
55 Kogure (2001)
56 Shield (1994)
57 Janet (1877) in James
   (1890)
58 Lemlich (1975)
59 Friedman et al. (2010)
60 See Friedman et al.
   (2010) again
61 Fradera & Ward (2006)
62 Linton (1975)
63 Walker (2003)
64 Ross & Wilson (2002)
65 Skowronski et al. (2003)
66 Wagenaar (1986)
67 Maycock et al. (1991)
68 Prohaska et al. (1998)
69 Frederickson et al. (2003)
70 Friedman (1987)
71 Crawley & Pring (2000)
72 Holmes & Conway
   (1999)
73 Conway & Haque (1999)
74 Linton (1988)
75 Frankl (1946)
76 Mann (1924). Quotes
   from p.104.

5. REMEMBERING THE FUTURE
77 D'Argembeau et al.
   (2011)
78 Rosenbaum et al. (2005)
79 Schacter & Addis (2007)

80 Hassabis & Maguire (2009)

81 Addis et al. (2008)

82 Kennett & Matthews (2009)

83 Eichenbaum & Fortin (2009)

84 Szpunar et al. (2007)

85 Hassabis et al. (2007)

86 Logan, C.J. et al. (2011)

87 Suddendorf & Corballis (2007)

88 Busby & Suddendorf (2005)

89 Atance (2008)

90 Buckner (2010)

91 Szpunar & McDermott (2008)

92 Berntsen & Bohn (2010)

93 Newby-Clark & Ross (2003)

94 Lachman et al. (2008)

95 Van Boven & Ashworth (2007)

96 Taylor et al. (1998)

97 Hawton (2005)

98 Crane et al. (2011)

99 Holmes et al. (2007)

100 Killingsworth & Gilbert (2010)

101 Bar (2009)

102 Azy et al. (2008)

103 See Hassabis & Maguire (2009) again

104 Gilbert & Wilson (2009)

105 Gilbert (2006)

106 Loewenstein & Frederick (1997)

107 Gilbert et al. (1998)

108 Wilson et al. (2000)

109 Dunn et al. (2003)

110 Liberman & Trope (1998)

111 Nussbaum et al. (2006)

112 See Liberman & Trope (1998) again

113 Wakslak et al. (2008)

114 Shu & Gneezy (2010)

115 Marshall (undated)

116 Buehler et al. (1994)

117 Mischel et al. (1989)

118 Steinberg ct al. (2009)

119 El Sawy (1983)

120 Weick (1995)

6. CHANGING YOUR RELATIONSHIP WITH TIME

121 Mangan et al. (1996)

122 O'Reilly (2000)

123 Tobin et al. (2010)

124 Schwartz (1975)

125 Hoffman (2009)

126 See Friedman et al. (2010) again

127 Ofcom (2010)

128 Bluedorn (2002)

129 Leroy (2009)

130 Jiga-Boy et al. (2010)

131 Putnam (1995)

132 See Frankl (1946) again

133 Kerkhof (2010)

134 James (1890)

135 Csikszentmihalyi (1996)

136 Zimbardo & Boyd (2008)

137 Williams & Penman (2011)

BIBLIOGRAPHY

This list is not exhaustive, but these are the main research papers to which I refer in *Time Warped*, and the books in this field that I found to be the most useful to my own research.

Apologies to the authors, but, to save space and trees, where there are multiple authors I've included only the first.

Addis, D.R. et al. (2008) Age-related changes in the episodic simulation of future events. *Psychological Science*, 19, 33–41.

Atance, C.M. (2008) Future thinking in young children. *Current Directions in Psychological Science*, 17, 2008, 295–298.

Azy, S. et al. (2008) Self in Time: Imagined self-location influences neural activity related to mental time travel. *Journal of Neuroscience*, 28(25), 6502–6507.

Baddeley, A.D. (1966) Time estimation at reduced body-temperature. *The American Journal of Psychology*, 79 (3), 475–479.

Baddeley, A.D et al. (2009) *Memory*. Hove: Psychology Press.

Bar, M. (2009) The proactive brain: memory for predictions.

Theme issue. Predictions in the brain: Using our past to generate a future. *Philosophical Transactions of the Royal Society*, B, 364, 1235–1243.

Bargh, J.A., Chen, M., & Burrows, L. et al. (1996) Automaticity of social behavior: Direct effects of trait construct and stereotype activation on action. *Journal of Personality and Social Psychology*, 71, 230–244.

Bar-Haim, Y. et al. (2010) When time slows down: The influence of threat on time perception in anxiety. *Cognition & Emotion*, 24(2), 255–263.

Berntsen, D. & Bohn, A. (2010) Remembering and forecasting. The relation between autobiographical memory and episodic future thinking. *Memory and Cognition*, 38(3), 265–278.

Bluedorn, A.C. (2002) *The Human Organization of Time: Temporal realities and experience*. USA: Stanford University Press.

Boring, L.D. & Boring, E.G. (1917). Temporal judgements after sleep. *Studies in Psychology*, Titchener Commemorative Volume, 255–279.

Boroditsky, L. (2000) Metaphoric structuring: Understanding time through spatial metaphors. *Cognition*, 75, 1–28.

Boroditsky, L. (2008) Do English and Mandarin speakers think differently about time? In B.C. Love et al. (eds) *Proceedings of the 30th Annual Conference of the Cognitive Science Society*, 64–70.

Boroditsky, L. & Ramscar, M. (2002) The roles of body and mind in abstract thought. *Psychological Science*, 13(2), 185–188.

Broome, M.R. (2005) Suffering and eternal recurrence of the same: The neuroscience, psychopathology and philosophy of time. *Philosophy, Psychiatry and Psychology*, 12, 187–194.

Broome, M.R. & Bortolotti, L. (eds) (2009) *Psychiatry as Cognitive Neuroscience*. Oxford: Oxford University Press.

Buckner, R. (2010) The role of the hippocampus in prediction and imagination. *Annual Review of Psychology*, 61, 27–48.

Buehler, R. et al. (1994) Exploring the "planning fallacy": Why people underestimate their task completion times. *Journal of Personality and Social Psychology*, 67(3), 366–381.

Buonomano, D.V. et al. (2009) Influence of the interstimulus interval on temporal processing and learning: Testing the state-dependent network model. *Philosophical Transactions of the Royal Society*, B, 364(1525), 1865–1873.

Busby, J. & Suddendorf, T. (2005) Recalling yesterday and predicting tomorrow. *Cognitive Development*, 20, 362–372.

Casasanto, D. (2010) 'Space for Thinking'. In Evans, V. & Chilton, P. (eds) *Language, Cognition and Space*. London: Equinox.

Casasanto, D. & Boroditsky, L. (2008) Time in the Mind: Using space to think about time. *Cognition*, 106, 579–593.

Casasanto, D. & Bottini, R. (2010) Can mirror-reading reverse the flow of time? *Spatial Cognition*, VII, 335–345.

Conway, M.A. & Haque, S. (1999) Overshadowing the reminiscence bump: Memories of a struggle for independence. *Journal of Adult Development*, 6, 35–44.

Cotard (1882) Du délire des negations. *Archives de neurologie*, 4, 152–170.

Cottle, T. (1976) *Perceiving Time: A psychological investigation with men and women*. New York: John Wiley & Sons.

Craig, A.D. (2009) Emotional moments across time: A possible neural basis for time perception in the anterior insula. *Philosophical Transactions of the Royal Society*, B, 364, 1933–42.

Crane, C. et al. (2011) Suicidal imagery in a previously depressed community sample. *Clinical Psychology & Psychotherapy* doi: 10.1002/cpp.741

Crawley, S.E. & Pring, L. (2000) When did Mrs Thatcher resign? The effects of ageing on the dating of public events. *Memory*, 8(2), 111–21.

Csikszentmihalyi, M. (1996) *Creativity: Flow and the psychology of discovery and invention*. New York: Harper Perennial.

D'Argembeau, A. et al. (2011) Frequency, characteristics and functions of future-oriented thoughts in daily life. *Applied Cognitive Psychology*, 25: 96–103.

Draaisma, D. (2006) *Why Life Speeds Up As You Get Older: How memory shapes our past*. Cambridge: Cambridge University Press.

Dunn, E.W. et al. (2003) Location, location, location: the misprediction of satisfaction in housing lotteries. *Personality & Social Psychology Bulletin*, 29(11), 1421–1432.

Eagleman, D.M. & Pariyadath, V. (2009) Is subjective duration a signature of coding efficiency? *Philosophical Transactions of the Royal Society*, B, 364(1525), 1841–1851.

Eichenbaum, H. & Fortin, N.J. (2009) The neurobiology of memory based predictions. *Philosophical Transactions of the Royal Society*, B, 364, 1183–1191.

El Sawy, O.A. (1983) Temporal perspective and managerial attention: A study of chief executive strategic behaviour. *Dissertation Abstracts International*, 44(05A), 1556–7.

Flaherty, M.G. (1998) *Notes on a Watched Pot*. New York: New York University Press.

Foster, R. & Kreitzman, L. (2003) *Rhythms of Life: The biological clocks that control the daily lives of every living thing*. London: Profile Books.

Fradera, A. & Ward, J. (2006) Placing events in time: the role of autobiographical recollection. *Memory*, 14(7), 834–845.

Frankl, V. (1946) *Man's Search for Meaning*. 2004 edition. London: Rider Books.

Frederickson, B.L. et al. (2003) What good are positive emotions in crises? A prospective study of resilience and emotions following the terrorist attacks on the United States on

September 11th, 2001. *Journal of Personality and Social Psychology*, 84(2), 365–376.

Friedman, W.J. (1987) A follow-up to 'scale effects in memory for the time of events': The earthquake study. *Memory and Cognition*, 15, 518–520.

Friedman, W.J. et al. (2010) Aging and the speed of time. *Acta Psychologica*, 134, 130–141.

Gevers, W. et al. (2003) The mental representation of ordinal sequences is spatially organized. *Cognition*, 87(3), B87–B95.

Gilbert, D.T. (2006) *Stumbling on Happiness*. London: Harper Press.

Gilbert, D.T. et al. (1998) Immune neglect: A source of durability bias in affective forecasting. *Journal of Personality & Social Psychology*, 75(3), 617–638.

Gilbert, D.T. & Wilson, D.W. (2009) Why the brain talks to itself: Sources of error in emotional prediction. *Philosophical Transactions of the Royal Society*, B, 364, 1335–1341.

Halberg, F. et al. (2008) Chronomics, human time estimation, and aging. *Clinical Interventions in Aging*, 3(4) 749–760.

Hassabis, D. & Maguire, E.A. (2009) The construction system of the brain. *Philosophical Transactions of the Royal Society*, B, 364, 1263–1271.

Hassabis, D. et al. (2007) Patients with hippocampal amnesia cannot imagine new experiences. *Proceedings of the National Association of Sciences*, 104, 1726–1731.

Hauser, D. J. et al. (2009) Mellow Monday and furious Friday: The approach-related link between anger and time representation. *Cognition and Emotion*, 23, 1166–1180.

Hawton, K. (2005) Restriction of access to methods of suicide as a means of suicide prevention. In Hawton, K. (ed.) *Prevention and Treatment of Suicidal Behaviour: From science to practice*. Oxford: Oxford University Press.

Henderson, J. et al. (2006) Timing in free-living rufous humming-birds, Selasphorus rufus. *Current Biology*, 16(5), 512–515.

Hoagland, H. (1933) The physiological control of judgments of duration: Evidence for a chemical clock. *Journal of General Psychology*, 9, 267–287.

Hoffman, E. (2009) *Time*. London: Profile.

Holmes, A. & Conway, M. A. (1999) Generation identity and the reminiscence bump: Memories for public and private events. *Journal of Adult Development*, 6(1) 21–34.

Holmes, E. et al. (2007) Imagery about suicide in depression – flash-forwards? *Journal of Behavior Therapy and Experimental Psychiatry*, 38(4), 423–434.

Hunt, A.R. (2008) Taking a long look at action and time perception. *Journal of Experimental Psychology, Human Perception and Performance*, 34(1) 125–136.

Husserl, E. (1893–1917) *On the Phenomenology of the Consciousness of Internal Time (1893–1917)*, translated (1990) by J.B. Brough. Dordrecht: Kluwer.

James, W. (1890) *The Principles of Psychology*. Vol 1, published 1907. London: Macmillan & Co.

Jiga-Boy, G.M. et al. (2010) So much to do and so little time: Effort and perceived temporal distance. *Psychological Science*, 21(12), 1811–1817.

Kennett, J. & Matthews, S. (2009) Mental time travel, agency and responsibility. In Broome, M. & Bortolotti, L. (eds) *Psychiatry as Cognitive Neuroscience: Philosophical perspectives*. Oxford: Oxford University Press.

Kerkhof, A. (2010) *Stop Worrying: Get your life back on track with CBT*. Berkshire: Open University Press.

Killingsworth, M.A. & Gilbert, D. (2010) A wandering mind is an unhappy mind. *Science*, 330, 932.

Klein, S. (2006) *Time: A user's guide*. London: Penguin

Koch, G. et al. (2002) Selective deficit of time perception in a patient with right prefrontal cortex lesion. *Neurology*, 59(10), 1658–1659.

Kogure, T. et al. (2001) Characteristics of proper names and temporal memory of social news events. *Memory*, 9(2), 103–16.

Lachman, M. et al. (2008) Realism and illusion in Americans' temporal views of their life satisfaction: Age differences in reconstructing the past and anticipating the future. *Psychological Science*, 9, 889–897.

Langer, E.J. (2009) *Counterclockwise*. New York: Ballantine Books.

Langer, J. et al. (1961) The effect of danger upon the experience of time. *The American Journal of Psychology*, 74(1), 94–97.

Leafhead, K.M. & Kopelman, M.D. (1999) 'Recent advances in moving backwards'. In Della Salla, S. (ed.) *Mind Myths*. New York: Wiley.

Lemlich, R. (1975) Subjective acceleration of time with aging. *Perceptual and Motor Skills*, 41, 235–238.

Leroy, S. (2009) Why is it so hard to do my work? The challenge of attention residue when switching between work tasks. *Organizational Behavior and Human Decision Processes*, 109, 168–181.

Levine, R. (2006) *A Geography of Time: The temporal misadventures of a social psychologist*. Oxford: Oneworld.

Lewis, P.A. & Miall, R.C. (2009) The precision of temporal judgement: milliseconds, many minutes, and beyond. *Philosophical Transactions of the Royal Society*, B, 364, 1897–1905.

Liberman, N. & Trope, Y. (1998) The role of feasibility and desirability considerations in near and distant future decisions: A test of temporal construal theory. *Journal of Personality and Social Psychology*, 75, 5–18.

Linton, M.R. (1988) 'Ways of searching and the contents of

memory'. In Rubin, M.R. (ed.) *Autobiographical Memory.* Cambridge University Press: Cambridge.

Linton, M. (1975) 'Take-two-items-a-day-for-five-years study'. In Norman, D.A. et al. (eds) *Explorations in Cognition.* W.H. Freeman: San Francisco.

Loewenstein, G. & Frederick, S. (1997) 'Predicting reactions to environmental change'. In Bazerman, H.H. et al. (eds.) *Environment, Ethics & Behaviour.* San Francisco: New Lexington.

Loftus, E.F. et al. (1987) Time went by so slowly: Overestimation of event duration by males and females. *Applied Cognitive Psychology,* 1, 3–13.

Logan, C.J. et al. (2011) A case of mental time travel in ant-following birds? *Behavioral Ecology,* 22(6), 1149–1153.

Mangan, P.A. et al. (1996) Altered time perception in elderly humans results from the slowing of an internal clock. *Society for Neuroscience Abstracts,* 22, 183.

Mann, H. et al. (2009) Time–space synaesthesia – a cognitive advantage? *Consciousness and Cognition,* 18, 619–627.

Mann, T. (1924) *The Magic Mountain.* Translated (1927) by Lowe-Porter, H.T. London: Vintage (1999).

Margulies, S.O. & Crawford, L.E. (2008) Event valence and spatial metaphors of time. *Cognition & Emotion,* 22(7), 1401–1414.

Marshall, F. (undated) History of the Philological society: The early years. http://www.philsoc.org.uk/history

Maycock, G. et al. (1991) The accident liability of car drivers. *TRL Research Report* 315. Berkshire: Transport Research Laboratory TRL.

McGaugh, J.L. (2003) *Memory and Emotion.* London: Weidenfeld & Nicolson.

McGrath, J.E. & Tschan, F. (2004) *Temporal Matters in Social Psychology.* Washington: American Psychological Association.

McNally, R.J. (2003) *Remembering Trauma.* Cambridge, Mass.: The Belknap Press.

McTaggart, J.E. (1908) The unreality of time. *Mind*, 17, 456–73.

Merritt, D.J. et al. (2010) Do monkeys think in metaphors? Representations of space and time in monkeys and humans. *Cognition*, 117, 191–202.

Miles, L. et al. (2010) Moving through time. *Psychological Science*, 21(2), 222–223.

Mischel, W. et al. (1989) Delay of gratification in children. *Science*, 244, 933–938.

Newby-Clark, I.R. & Ross, M. (2003) Conceiving the past and future. *Personality and Social Psychology Bulletin*, 29, 807–818.

Noulhiane, M. et al. (2007) How emotional auditory stimuli modulate time perception. *Emotion*, 7(4), 697–704.

Nussbaum, S. et al. (2006) Predicting the near and distant future. *Journal of Experimental Psychology*, 135, 152–161.

O'Reilly, K. (2000) *The British in the Costa del Sol*. London: Routledge

Ofcom (2010) *The Communications Market* 2010. London: Ofcom.

Pöppel, E.(2009) Pre-semantically defined temporal windows for cognitive processing. *Philosophical Transactions of the Royal Society*, B, 364, 1887–1896.

Prohaska, V. et al. (1998) Forward telescoping: the question matters. *Memory*, 6, 455–465.

Putnam, R. (1995) Bowling alone: America's declining social capital. *Journal of Democracy*, 6(1), 65–78.

Rosenbaum et al.. (2005) The case of K.C.: Contributions of a memory-impaired person to memory theory. *Neuropsychologia*, 43, 989–1021.

Ross, M. & Wilson, A.E. (2002) It feels like yesterday: Self-esteem, valence of personal past experiences, and judgments of subjective distance. *Journal of Personality and Social Psychology*, 2002, 82(5), 792–803.

Schacter, D. (1996) *Searching for Memory*. New York: Basic Books.

Schacter, D.L. & Addis, D.R. (2007) The cognitive neuroscience of constructive memory: Remembering the past and imagining the future. *Philosophical Transactions of the Royal Society*, B, 362, 773–786.

Schleidt, M.M. & Eibesfeldt, E. (1987) A universal constant in temporal segmentation of human short-term behaviour. *Naturwissenschaften*, 74, 289–290.

Schwartz, B. (1975) *Queuing and Waiting*. Chicago: University of Chicago.

Sevinc, E. (2007) The effects of extensive musical training on time perception regarding hemispheric lateralization, different time ranges and generalization to different modalities. PhD thesis retrieved 10.01.12. from http://cogprints.org/6171

Shield, R. (1994) Extracts from diary. *NPR Sound portraits*. http://www.soundportraits.org/on-air/worlds_longest_diary/diary_entry1.gif

Shneidman, E.S. (1973) Suicide notes reconsidered. *Psychiatry*, 36, 379–393.

Shu, S.B. & Gneezy, A. (2010) Procrastination of enjoyable experiences. *Journal of Marketing Research*, 47(5) 933–934.

Siffre, M. (1965) *Beyond Time*. London: Chatto & Windus.

Skowronski, J.J. et al. (2003) Ordering our world: An examination of time in autobiographical memory. *Memory*, 11, 247–260.

St Augustine (2004) *Confessions of a Sinner*. London: Penguin

Steinberg, L. et al. (2009) Age differences in future orientation and delay discounting. *Child Development*, 80(1), 28–44.

Suddendorf, T. & Corballis, M.C. (2007) The evolution of foresight: What is mental time travel and is it unique to humans? *Behavioral and Brain Sciences*, 30, 299–313.

Szpunar, K.K. & McDermott, K.B. (2008) Episodic future

thought and its relation to remembering: Evidence from ratings of subjective experience. *Consciousness and Cognition*, 17, 330–334.

Szpunar, K.K. et al. (2007) Neural substrates of envisioning the future. *Proceedings of the National Academy of Sciences*, 104, 642–647.

Taylor, S. (2007) *Making Time:Why time seems to pass at different speeds and how to control it*. London: Icon Books.

Taylor, S.E. et al. (1998) Harnessing the imagination: Mental simulation, self-regulation and coping. *American Psychologist*, 53, 429–439.

Tobin, S. et al. (2010) An ecological approach to prospective and retrospective timing on long durations: A study involving gamers. *PLoS One*, 5(2), e9271.

Twenge, J.M. et al. (2003) Social exclusion and the deconstructed state: Time perception, meaningless, lethargy, lack of emotion, and self-awareness. *Journal of Personality and Social Psychology*, 85(3), 409–423.

Van Boven, L. & Ashworth, L. (2007) Looking forward, looking back: Anticipation is more evocative than retrospection. *Journal of Experimental Psychology: General*, 136, 289–300.

van Wassenhove, V. (2009) Minding time in an amodal representational space. *Philosophical Transactions of The Royal Society*, B, 364, 1815–1830.

Vicario, C. et al. (2010) Time processing in children with Tourette's syndrome. *Brain and Cognition*, 73 (1), 28–34.

Wagenaar, W.A. (1986) My memory: A study of autobiographical memory over six years. *Cognitive Psychology*, 18, 225–52.

Wakslak, C.J. et al. (2008) Representations of the self in the near and distant future. *Journal of Personality and Social Psychology*, 95, 757–773.

Walker et al. (2003) Life is pleasant – and memory helps to keep it that way. *Review of General Psychology*, 7(2), 203–210.

Ward, J. (2008) *The Frog who Croaked Blue*. London: Routledge.

Weick, K.E. (1995) *Sensemaking in Organizations*. California: Sage Publications.

Williams, M. & Penman, D. (2011) *Mindfulness: A practical guide to finding peace in a frantic world*. London: Piatkus

Wilson, T. et al. (2000) Focalism: A source of durability bias in affective forecasting. *Journal of Personality and Social Psychology*, 78(5), 821–836.

Wittman, M. (2009) The inner experience of time. *Philosophical Transactions of the Royal Society*, B, 364, 1955–1967.

Wyllie, M. (2005) Lived time and psychopathology. *Philosophy, Psychiatry & Psychology*, 12 (3), 173–185.

Zakay, D. & Block, R.A. (1997) Temporal Cognition. *Current Directions in Psychological Science*, 6, 12–16.

Zerubavel, E. (2003) *Time Maps: Collective memory and the shape of the social past*. Chicago: University of Chicago Press.

Zerubavel, E. (1981) *Hidden Rhythms: Schedules and calendars in social life*. Chicago: University of Chicago Press.

Zhong, C. & DeVoe, S. (2010) You are how you eat: Fast food and impatience. *Psychological Science*, 21, 619–622.

Zimbardo, P. & Boyd, J. (2008) *The Time Paradox*. London: Rider Books.

INDEX